Praise for UNZIP YOUR GENES

"Fascinating, eye-opening, informative! *Unzip Your Genes* is a must-read for anyone looking to live a happier, healthier life. Dr. Jennifer Stagg takes her expertise in precision lifestyle medicine to a new level by allowing readers to understand how their bodies truly work from the inside out."

—*Keisha Grant*, TV News Anchor

"Your genetic make-up not only defines who you are, but how you interact and react to your environment. Dr. Stagg's new book captures in a lucid and compelling manner, what so many before her have been unable to do: Explaining why genetics has an important role in the everyday discourse of health and wellbeing, while striking an admirable balance between simplicity and scientific accuracy. Dr Stagg introduces genetics in a manner that is so much more relevant than just disease diagnostics—genetics that helps us to better understand ourselves as human beings. I highly recommend Dr. Stagg's *Unzip your Genes* for anyone interested in the dynamic and interactive role that individualized genetics play in one's health and wellbeing (within the framework of one's environment). I suspect that after reading her book, many will be inspired to begin their own personalized journey to optimized health."

—*Dr. Mansoor Mohammed, PhD*,
Founder and President of ManageneDx, Innovative Lifestyle
Genomics, Health-focused Personalized Medicine

"In the confusing and often daunting world of nutrigenomics, Jennifer Stagg has come up with a guide to understand and conquer this arena in practical and effective ways. *Unzip Your Genes* is a wonderful start to a whole new you by unlocking your potential."

—*Christopher Keroack, MD*, Author, Speaker and Medical
Consultant in Lifestyle and Functional Medicine

T0019481

"Every doctor advises their patients to eat better, exercise more and stress less. As a result, we have more gyms than ever, we spend more on diets than ever, and yet we have a runaway epidemic of obesity and diabetes which spurs the top two killers in the country, namely cardiovascular disease and cancer. So the question is, why are these simple strategies are not working?

The answers are uncovered within the pages of *Unzip Your Genes*. Dr Stagg eloquently lays out the argument for the necessity for targeted, true Health Care. Or as she describes, 'Precision lifestyle medicine.'

This book should be required reading for any health care provider who is not just interesting in defeating disease, but in the promotion of wellness!

Good work Dr. Stagg, I enthusiastically and emphatically endorse your book!"

—*Dr. Douglas A. DiSiena, DC, FICA, CNS*, QME Distinguished Fellow of the ICA Bestselling co-author of Possibility Living

"Dr. Stagg wrote a comprehensive book that is an essential read for every generation! She created an amazing precision program to truly unlocking your health and creating a lifetime of wellness."

—*Dr. Pina LoGiudice, ND, LAc*, Author of The Little Book of Healthy Beauty

"At a time that we are so confused about how to be healthy, Dr. Jennifer Stagg makes it so simple and practical with her new book, *Unzip Your Genes*. This is a must read for everyone!"

—*Dr. Fabrizio Mancini*, International Speaker and bestselling author of The Power of Self-Healing

UNZIP *your* GENES

UNZIP *your* GENES

5 Choices To Reveal A Radically Radiant You

DR. JENNIFER STAGG

A SAVIO REPUBLIC BOOK
An Imprint of Post Hill Press
ISBN: 978-1-68261-644-4

Unzip Your Genes:
5 Choices to Reveal a Radically Radiant You
© 2017 by Jennifer Stagg
All Rights Reserved
First Savio Republic Hardcover Edition: November 2016

No part of this book may be reproduced, stored in a retrieval system, or transmitted by any means without the written permission of the author and publisher.

posthillpress.com
New York • Nashville
Published in the United States of America

Contents

Part 2 EPIGENETICS: How Your DNA is Affected by Your Environment

Part 3   THE PRECISION HEALTH PROGRAM

Introduction

Technological advances have dramatically impacted almost every aspect of daily life, from how we communicate to how we buy things to how we bank. Now, there's a high-tech and scientific revolution underway in health care that can transform patient care. I'm not referring to a new treatment for one disease but rather something that can help both physicians and patients.

Genomics. If you're a science geek or you remember your Biology 101, you know that our bodies contain DNA, which is the so-called building block of life. Genomics goes beyond genetics. The genome is the entire DNA content in one cell of an organism. Scientists study the complete DNA sequences and perform genetic mapping to help understand what causes disease. Genomics allows us to see how variations in our DNA can interact with one another and impact our growth, development and overall health. You've probably heard of genetic testing for the BRCA gene which can cause breast cancer. While it's true that genomics allows people to find out if they are genetically predisposed to a particular disease, there's much more to this science. By understanding your unique gene structure, you can view your body in an innovative way that allows you to take control of your health and your future.

You should be reassured by the following two facts:

1. There are more genes associated with resilience and the ability to keep your body in balance or homeostasis than there are for disease.
2. Your genes are not your destiny. The way your genes get expressed is affected by what we call "genetic modifiers," including your lifestyle and environmental and psychosocial factors.

Each of you is unique and there are specific actionable steps you can take that can lead to wellness. With guided personalized help, or what I like to call "precision prevention," you can get the most out of your genes.

For me, after practicing medicine for 15 years, providing patients with individualized plans to meet their nutritional needs based on the latest research combined with clinical data, genomics has allowed me to take a quantum leap. I can custom design my patient's lifestyle medicine program by utilizing genetic testing and other personalized evaluation tools that reveal each person's "unique code." This is a cutting-edge field of health care that I refer to as Precision Lifestyle Medicine. As part of this new type of medicine, you-the-patient become an active participant in unlocking your perfect combination, or "code," of practices for your body, mind, and spirit. I'm not simply treating you for high cholesterol or weight management. I look at every aspect of your body to help you make changes so you can experience a fuller, happier, and more rewarding life. If you want to feel *your* best, function at *your* highest level, and live *your* life to the fullest extent, this book is for you!

The patients in my practice vary widely in terms of their wellness. Unfortunately, over the years, I see more and more patients who have multiple chronic health problems including heart disease, mood disorders, hormonal imbalance, gastrointestinal complaints, and obesity. As a physician trained to understand and treat the root cause of illness and identify what is preventing someone from being healthy, I always examine the link between a patient's lifestyle and his or her unique body chemistry.

My approach differs dramatically from that of many of my peers in the medical community who follow the traditional disease-focused model of health care. In fact, I believe that the healthcare crisis we are facing in this country is the result of concentrating on the treatment of disease states, rather than what makes us healthy. This, in my opinion, is what constitutes true health care. For the first time, we are seeing scientists studying the genes that can keep people healthy, even when they are exposed to a host of environmental and genetic factors that would make someone else quite sick. This is called the study of resilience. By cataloging the DNA of a large number of people, scientists can identify these "resilience genes" and through extensive study of the subjects' genetics, epigenetics and metabolism, help us understand why some people overcome an illness or remain healthy while others get sick.

For the majority of my patients, by using advanced clinical diagnostic skills and relying on proven nutritional and lifestyle strategies, we can, over time, restore them to a state of health and help them achieve balance. However, for a subset of patients, this plan of treatment just doesn't work. These patients appear to be doing everything right. They are compliant with treatment protocols and follow the outlined program, but it produces minimal results or just doesn't work at all. This "failed treatment" occurs more often with patients (especially women) who are overweight or obese and suffer from many chronic health problems. Sound familiar?

I commonly hear this complaint from patients in my office and others in social settings (especially when people find out what I do for a living): "My friends eat the same way as I do and they are not overweight. I promise, I don't go home and eat half a chocolate cake at night. I just don't understand why I can't lose this weight?"

I sometimes feel like a detective trying to solve a mystery. I begin with a comprehensive evaluation which starts the moment I meet patients. I take a thorough medical and family history, finding out what previous treatments worked or failed, to get clues to the underlying obstacles preventing them from their true state of health. I then move on to laboratory assessments including basic tests for metabolic function, lipids, insulin sensitivity, hemoglobin A1C, and hormones (thyroid, adrenal, sex). Investigating digestive function, detoxification, cellular metabolism, and neurotransmitters is an additional requirement for some people. And, of course, now I utilize genomic testing.

Genomic testing provides information that was not clinically available just a few years ago. I can now reply to patient questions such as, "It seems like I get fat when I weight-train. Is that possible?" or "I've tried every low-carb diet and it doesn't work. How could that be?" It is really quite simple. A patient's saliva sample can provide answers. "Mrs. Smith, genetically you may be more likely to gain fat mass with an intensive strength training program" and "Ms. Morgan, with your genetic makeup, you will do better on a Mediterranean diet."

But, as we all know, there is more to a healthy, well-balanced life than diet and exercise. I have been a practicing physician long enough to a have gained a deep understanding of the connection of mind, body, and spirit—all of which impact our DNA. Clinically, I see patients who are following the right diet and exercise program, practicing stress reduction, yoga, and socializing, but there is still a missing connection.

One patient was feeling unbalanced about her children leaving for college. Some physicians might say she was suffering from "empty nest syndrome" and perhaps refer her to counseling and advise her to pick up a hobby or start volunteering. She had been a stay-at-home mother and even some friends were criticizing her, wondering how she would keep busy without her children at home. She was feeling overwhelmed and even starting to question her purpose in life. I simply asked her, "What if your purpose in life is just to be happy?" She seemed intrigued and left the office saying she was going to spend some time thinking about that. Two months later, she returned to my office and appeared completely different. She told me that what I had said at her last visit had changed her life. She also explained that she felt she had experienced a spiritual awakening. She approached each day, and even each moment in life, from a new perspective. She now asked herself, "Does this make me happy?" and made decisions from that space. This is a clear example of the complex interplay of mind, body, and spirit that impacts function and creates a state of wellness and vitality.

You may have noticed that some people have figured out, on their own, how to achieve the perfect balance of health and well-being (and those people usually don't present in my office!). My patients often ask what I eat, what type of exercise I do, how I manage stress, and even which products I use on my skin! People are intrigued by what makes others healthy and mistakenly think that what works for one person could work for them. This is the essence of genomics—you are unique and what allows you to manifest your true state of health is also unique. So, yes, personally, as a happily married working mother of three, who has been searching for the "code" since I began my professional career more than 20 years ago, I thought I had figured it out for myself through a combination of educated guess and trial and error. But, by applying the information I discovered about my genomics, I feel I have finally deciphered *my* code for a healthy and, most importantly, happy life. In this book, I will help you do the same—but I'll save you the two decades of discovery that it took me!

I know you want to find out the important stuff—how to get the body you always wanted, how to sleep longer, how not to feel stressed all the time and more. I will discuss these issues, but I believe it's very important you understand the basics of genomics. Even if you hated high school biology class, I encourage you to read about the scientific studies.

Knowing what your genetic makeup reveals about your body provides added motivation to make, as well as maintain, lifestyle changes. That is the key to becoming healthier.

This book is divided into three parts. Part 1 introduces you to the basics of genomics and the five areas that impact your health. Part 2 examines epigenetics, the science of nurturing and how your DNA interacts with the environment. Part 3 is the Precision Health Program which focuses on improving your mind, body and spirit, according to your individual genetic profile.

I would never suggest to you that I have all the answers. In fact, I can say confidently that I do not know it all. This field and our knowledge about human health and wellness, in general, are still expanding. And, as I learn more, sometimes I find that the way I thought things worked is actually not the case. Being open to new ideas and having a willingness to adopt novel concepts in light of compelling evidence is a fundamental value I committed to a long time ago. Incorporating DNA testing and applying the results in a lifestyle approach is a huge paradigm shift in the healthcare industry. A more significant shift is the concept of *promoting* health and empowering individuals, instead of the disease treatment model.

I feel we are really just at the tip of the iceberg in the transformation of health care and wellness. The key to moving forward is continuing to assimilate the constant flow of incoming information, while embracing our own unique potential and personal power. I believe that is the pathway to health, abundance, and prosperity.

Part 1

GENOMICS

How DNA Influences Your Health

CHAPTER 1

The One Thing that Could Improve Your Health and Longevity

For years, when asked to name the one strategy that would improve a person's health, my response would be related to diet. I would advise the person to cut back on sugar, eliminate processed foods, and follow a Mediterranean-style diet. Then, I might start explaining the importance of being active and having a positive attitude. Today, however, my response would be very different. It's not that my earlier advice was wrong but the reality is that technology has given physicians—and in turn patients—a far more powerful arsenal. Precision medicine and more specifically genetic testing is the newest way to understand your unique needs related to nutrition, exercise, and risks of metabolic conditions including diabetes, high cholesterol, and obesity. Reviewing your test results with a

1

healthcare provider is referred to as genetic counseling and understanding your profile increases the chances you'll be able to successfully transform your lifestyle into a healthier one.

You might be inclined to think that if you found out you were genetically programmed for obesity you would probably just give up. In fact, the opposite is true! You don't just throw in the towel and say you are doomed. Research shows that when patients find out they have the genes for obesity, they actually are more likely to make healthier lifestyle choices than if they had not received genetic counseling.

In fact, one of the biggest challenges in private practice is motivating patients to make lasting changes that positively influence their overall health. In the healthcare field, we often refer to this as patient compliance. For some people, the most difficult part is getting started while, for others, it's sticking to those healthier behaviors over the long-term, choosing to make exercise a priority and making sure they fill their kitchen with smart food options. With three growing kids of my own, I know how challenging it can be to keep the fridge and pantry stocked, but I have decided that this is a must for me.

As a physician, I spend most of my day telling patients what they SHOULD do, but *you* have to do the real work after you leave your doctor's office. Fortunately, genetic counseling helps and can make your work, if not easier, more focused. Many of my patients view the genetic counseling process as therapeutic. Some people have said they felt lighter, relieved, and even inspired.

This type of genetic profiling also provides valuable information about eating behaviors so we can further boost the likelihood someone will be successful with a healthy eating program. There are six behavior traits that can be evaluated and based on the results, we have developed a tool to allow us to provide targeted behavioral counseling. For example, if a patient displayed a variant associated with eating *disinhibition*, essentially a lack of restraint, then we would advise the person to measure his or her food and not eat from an open bag. Also, to avoid overeating, the person should have nutritious food available, especially when facing an emotional stressor.

When reviewing their genetic profiles, I often tell my patients that genetics (or nature) is about 30% of the equation, which is great news because that means 70% is environmental (or nurture), which is what can be influenced by your lifestyle choices. How you respond to emotional stressors, food signals, and activity can all influence what happens to you. This is why a person can have the genes for obesity or diabetes but never actually become obese or get diagnosed with diabetes.

Another critical discovery is that genes can be "turned on" and "turned off." This is a branch of genetics referred to as epigenetics. This is a very active area of research and scientists have already uncovered fascinating data. For example, while it has been generally accepted that green tea is healthy, specific studies are now examining how phytochemicals in the tea affect your genes.

There is no question that precision medicine will change the face of health care, especially in the area of targeted pharmaceutical therapies. But my emphasis has always been on prevention and lifestyle medicine. I coined the term

"Precision Lifestyle Medicine" to specifically focus on the 70% of environmental factors that you can influence, once you are armed with your 30% genetic knowledge. While you can certainly benefit from the information in this book without getting a genetic test, in fact, this type of testing is widely available; and once you are armed with information on your genes, you'll be able to recognize the environmental cues that influence your genes in a way that will help you lead a healthier, happier and longer life.

The Polygenetic Theory of Obesity

This book is not a weight loss guide; my concern is improving the overall state of your health. However, because the most significant contributing factor to chronic disease is obesity, I will explain how to achieve and maintain a healthy body composition. The rates of obesity in the Western world have risen to epidemic levels. Addressing obesity has become an entire industry; the medical specialty on treating obesity is called bariatric medicine. The quest for a cure for obesity has continued to elude researchers. The most likely reason is that they are looking for the silver bullet—a solution in a pill, procedure or even gene therapy. But the most effective solution we have always had and will likely continue to have is lifestyle medicine—and it works!! For some people, though, the road is not as easy as it is for others.

There are many factors that play a role in the unique inter-actions between genetics and environment. Researchers have been examining genetic factors for many years and recent work in this field is the most promising to date. Scientists have shown that obesity can be attributed to a single gene variant only in a small percentage of people. For most people, their weight is a complex interplay of their unique genome and the lifestyle choices that affect how those genes are expressed. The concept that many genes are involved in obesity is called the Polygenetic Theory of Obesity. The majority of people who have problems with weight management likely have multiple genetic variants that are associated with obesity, metabolic factors and abnormal eating patterns. These are the reasons why managing obesity has been challenging, and the most effective treatment requires a highly individualized approach. What works for one person is unlikely to work for someone else because of the differences in DNA.

CHAPTER 2

The Language
of Genomics

Some of you may be intimidated by the word genomics but you shouldn't be; think about how quickly we've all become accustomed to the latest technological jargon. And usually a 10-year-old knows the terms before we've even heard of them! The reality is that our world is rapidly changing and the future of health care will be based on a combination of genomics and technology. Once you are familiar with the basic definitions, you'll be able to understand the rest of the concepts that I'll be discussing throughout this book. I'll walk you through the "ABCs" as painlessly as I can. And, if you want to learn more about the science of genomics, you can check out resources such as the National Human Genomic Research Institute https://www.genome.gov/ or the National Institutes of Health http://www.nih.gov/.

What is a SNP?

Although it may seem counterintuitive, I'm first going to explain what a SNP is before defining other genetic terminology. You will see the term SNP (pronounced "snip") repeatedly in this book and in the field of personalized medicine, in general. (If you are familiar with the fundamentals of genetics, then you can skip the next section.)

You have probably all heard of Wikipedia, but did you know there is a SNPedia? It is an entire web-based database that you can use to look up any SNP. Now what exactly is a SNP? A SNP is a single nucleotide polymorphism, to be exact. I know, that probably doesn't mean anything to you. To understand SNPs, you need to know a few key principles:

▸ Your genetic code is built with DNA, which is short for deoxyribonucleic acid.

▸ The smallest building blocks of DNA are called nucleotides.

▸ A nucleotide contains a base attached to a sugar molecule and a phosphate molecule.

▸ There are four bases: adenine (A), guanine (G), cytosine (C) and thymine (T).

▸ These bases come together as pairs: Adenine and guanine, AG; Cytosine and Thymine, CT

▸ The order of the bases is what is referred to as the sequence. The sequence is like our genetic alphabet.

▸ The nucleotides run along two strands that come together to form the familiar double helix of DNA.

As humans, about 99% of our genetic code is the same in each person. The most common variation that differentiates us is a SNP. This is where one nucleotide is replaced with another; for example, adenine (A) is replaced with guanine (G). There are approximately 10 million SNPS in the human genome. Usually the SNPs occur along a sequence of DNA, between sections that code for a gene, and this is not thought to have any health consequences. For example, this type of polymorphism is part of a complex interplay of many genes that is responsible for eye color. It is when the SNP occurs in a gene that we may see impacts on human health.

If you have heard about the concept of genetic variants, sometimes referred to as mutations, then you may have some familiarity with the SNPs that occur in the MTHFR (methyl tetrahydrofolate reductase) gene which affect the activity of that enzyme. These SNPs determine how well folic acid is converted to the active form methyl folate. People will take L-5-MTHF (L-5-methyltetrahydrofolate) in supplement form; there are even prescription medications to improve the effectiveness of antidepressant medications for patients even if they have not had genetic testing.

The Language of Genomics

Now, stick with me a little longer. If you've studied biology or genetics at all, you can skip to the next chapter. But most of you should acquaint with the basics in the next section.

We are all born with 23 pairs of **chromosomes** inside the **nucleus** of our cells. The nucleus can be thought of as the organizing center of the cell. Chromosomes contain our genetic material, or DNA, which we inherited from our parents: one set of 23 from our mothers and another set of 23 from our fathers. Your **genes** are distinct sequences along a section of DNA that provide the instructions to make proteins. DNA found in the nucleus of cells goes through a repeated process where it gets **transcribed** to eventually become a series of **amino acids**, which are the building blocks of **protein**. These proteins in your body have many critical roles, such as providing the structural components of cells, acting as enzymes that carry out complex biochemical reactions, transporting and storing smaller compounds and serving as antibodies that arm your immune system, and hormones that carry chemical messages. You may read that a gene "codes for" a certain protein, or enzyme. This just means that that small section of DNA can be transcribed as that protein.

Mutations in a gene are permanent changes or alterations in the DNA sequence. Gene mutations can either be inherited from your parents, which is referred to as **hereditary**, appearing in every cell in your body, or they can be **acquired** (or **somatic**) at some point during your life, often the result of interaction with your environment, and they do not appear in all cells. Most mutations don't result in any major problems. Also, when we talk about gene mutations or variations like SNPs, we also look to see whether there are one or two "copies" of that SNP, which can often impact the outcome that results in terms of physiologic function (also known as **phenotype**). If you have two copies of a SNP, then you are **homozygous**. If you only have one copy of a particular SNP, then you are **heterozygous**. In some cases, being heterozygous may still cause significant dysfunction to occur.

But usually, it means there will be a lesser effect on physiologic function. When you learn about your own genetics it is important to distinguish how the pattern of the SNP you carry could be affecting you.

Obviously, there's much more detailed intelligence on genomics that I could include here. But that's not the focus of this book. I want you to understand the many genetic variants that could be impacting your level of health. I'm focusing on identifying the SNPs for which you can take a series of actionable steps that help move you in the direction of vitality. The aim of this book is for you to do things that help you feel better NOW, not in 10 to 15 years. Keep reading and you'll be learning about this burgeoning field, until recently reserved to the elite. You'll become knowledgeable about this cutting-edge transformation in health care.

CHAPTER 3

#1 BEHAVIOR: Your Genes Dictate the Psychology of Eating

For years, scientists have suspected there are genetic reasons why eating patterns vary so much but proving the genetic link to obesity was difficult. Scientists were often baffled by data which didn't support their theory that certain populations would behave similarly. Today, with the remarkable advancements made in the field of genomics, we now have a greater understanding of human food seeking and eating behaviors that differ so widely in our population.

Nearly 20 years ago, a study in the *American Journal of Clinical Nutrition* demonstrated that a population of obesity-prone individuals preferred sweet and creamy foods less than the non-obese group, but any elevated response was later correlated to weight gain. Given that about 60% of the calories in the typical American diet came from refined sugars and fat, and prior research had linked high intakes of fat and sugar to elevated body mass index (BMI), this study focused on possible genetic reasons that would explain why some people consume these foods.

The hypothesis of this study was that taste preferences may be genetically encoded through taste receptors. Researchers chose male and female Pima Indians who were free of type 2 diabetes; as a group, these people are prone to obesity. Initially, the subjects were presented with a taste test of solutions varying in sugar and fat content. They rated these solutions for sweetness, creaminess, and pleasantness. Body measurements were also taken. Then the subjects returned after three months for a repeat taste test and their body measurements were also recorded. Body measurements were repeated in some subjects for periods as long as eight years. The same protocol was conducted with Caucasian subjects, a population not prone to obesity, for comparison.

The results showed that race was the only predictor of hedonic response, which is a rating of how pleasant the solution tasted. The Pima Indians had a lower pleasant response to the sweet and creamy solutions, but thought they tasted sweeter and creamier. Although the lower hedonic response contradicted their expectations, the researchers said this effect may have been due to an impaired pleasure response system in the brain, or possibly impaired insulin sensitivity. However, any heightened response to these tastes was correlated with weight gain in the Pima Indians. The scientists said that the finding "lends support to the hypothesis that a preference for highly palatable foods is associated with the development of obesity." Fast-forward a decade and we now know there is a genetic variant for preference for sweet tasting foods (which I will discuss in this chapter) and, of course, obesity genes, (which will be covered in chapter 5).

Eating Disinhibition—Hey, Put Down that Big Bag of Chips

✱ TAS2R38-rs1726866

❩This is the name of the SNP which will appear as part of the headings throughout section 1 as I discuss the most significant SNPs.)

Some of us know (even if we won't freely admit it) that we can't sit down with a full bag of chips, a tray of cookies, an entire dark chocolate bar...or whatever our vice may be. So let me state my ongoing mantra, **"There's a SNP for that!"** One of the primary reasons many people have difficulty maintaining healthy eating patterns is because of an eating behavior called **eating disinhibition**. This behavior can be explained by a variant within one of our genes. These variants are what make us unique. To be specific, this SNP is a variant in a gene related to eating behavior, called TAS2R38. And to make matters worse, it only applies to women! A study conducted in 2010 showed that women with the T allele of rs1726866 were more likely to display eating disinhibition in response to a stimulus. The stimulus was defined as something tasting really good (okay, that's trouble), situations that can trigger overeating, like certain social settings, or the classic, emotional upset. If you have ever uttered these words, "I'm an emotional eater," then you probably have this gene.

I know for many of the women I see in my practice, emotional eating or overeating carries a sense of shame, guilt, defeat, or failure. They tell me they don't understand why they continue to sabotage themselves. They know better, but it is so hard to control. They feel like they are purposely harming themselves; they know it makes no sense, but they keep doing it.

When I review their genomic profile related to eating behaviors and explain the biological reason for their behavior, many women feel a sense of relief, knowing that it's not their fault. They finally understand, after many years, why they struggled with their relationship with food while their friends did not suffer from the same problems. What I love about genomics is that it can provide a sense of empowerment. This is really just a variant in our human race. It's what makes us all unique! Embracing this knowledge is the first step. The next step is what I call **Precision Lifestyle Medicine**. We take this profile and specifically target the variation in eating behavior. This is not a one-size-fits-all approach. Many, many women are using behavior modification techniques that may offer them only minimal success because they are not addressing the unique nature of their true self. A woman with the SNP for eating disinhibition will do very poorly with the three-bite dessert rule! Instead, if you have the eating disinhibition SNP, you should follow specific targeted behaviors. I will discuss several strategies to use if you have this SNP in part 3 The Precision Health Program. For now, let me offer one obvious but very effective tip:

> ▸ **Recognize your triggers** for emotional eating and when you see one coming, have a substitute lined up. Find something other than food—music, exercise, dancing, talking to that friend who always makes you laugh!

Food Desire: Your First True Love?

✳ ANKK1/DRD2-rs1800497

Okay, I am being a little dramatic but our relationship with food is a huge part of our lives. Have you ever noticed that you seem to think about food a lot more than most of your friends? Occasionally, one of my kids will ask me at breakfast, "What are we going to have for dinner tonight?" Some of us are thinking ahead to what our next snack or meal will be while we're still in the middle of eating our current meal. This tendency is related to our general desire for food, which is extremely complex and influenced by a variety of factors.

In a study evaluating eating behaviors in a group of obese subjects related to food reinforcement in 2007, a SNP was identified that essentially demonstrates how much effort someone is willing to exert to obtain a specific food. When I am explaining this SNP to my patients, I often tell them, "This is the SNP that drives you to change out of your pajamas (or maybe not), get into your car and head off to the grocery store to get that pint of ice cream." It is similar to those intense pregnancy cravings when women must have a particular food, no matter what time it is or how odd the craving is.

So here are the technical details that I provide when I'm discussing SNPs: people who have this variant have the T allele in the rs1800497 genetic marker on the ANKK1/DRD2 gene. The normal genotype is reported as C/C. We label this SNP as FOOD DESIRE. In a study, subjects were asked to complete a task in exchange for a small amount of their favorite foods. The experiment proceeds to offer more and more difficult tasks until the subject decides the task is not worth the food reward and quits. This is a classic example of the pain and pleasure model. The early quitters are considered to have low food reinforcement. The subjects with the SNP were more likely to exert more effort to obtain their favorite foods (thus showing high food reinforcement), and eat more of them! Eureka!!

Now that eating behavior has been identified and labeled, what can you do if you have this SNP? The first step is recognizing that this is how you are programmed. It doesn't mean you're doomed to make endless late night trips to the grocery store. It does mean you need to bring in the heavy "reinforcements." Again, get the people you live with, whether they are roommates, spouse, parents, even your kids, involved. (If you live alone then start with the techniques below, but at least tell your loved ones. Call them when you begin thinking about heading out on that snack run.)

I love to frame it as being "blessed as you," or "what makes you special"!! Put a positive spin on it. Think of when you're in a serious relationship; you get the whole package. And the most important relationship you have is the one with yourself. Stop feeling guilty about your behaviors and shaming yourself. You have this one body and you can't trade it in! Embrace it. A dear friend of mine says that we should treat ourselves more like we would treat our sweet little children when they are sad or hurting. I just love that!

I'll include some helpful precision lifestyle strategies to address this food desire SNP in part 3. For now, the most important step you can take is making it more difficult to obtain your favorite food. There will come a point where the effort to obtain that food is just not worth it. You have to figure out what your threshold is, from driving to the store in your PJs late at night or having to go to another part of your home where you store the snacks.

Sweet Tooth, Not a Figment of Your Imagination

✳ SLC2A2-rs5400

Have you always had a so-called sweet tooth? Of course, we know that for some of us, the more we eat sugar, the more we want it. In fact, you may have heard that sugar is about 10 times more addictive than cocaine!! In fact, so many of my patients tell me they are addicted to sugar that I developed an intensive 28-day outpatient program in my clinic. The foundation of the program is a 28-day metabolic detoxification, intensive acupuncture treatment (which is used in many inpatient drug addiction centers) plus behavior modification therapy. After the initial stage, these patients transition to maintenance care, where the focus is continued behavior modification.

The behavior modification piece is of vital importance in patients who are genetically predisposed to addiction in general, but now we know there are also SNPs in our genes that affect our taste for sweets. There is actually a trait known as "sweet tooth" which is coded for by the rs5400 SNP and appears on chromosome 3 in the SLC2A2 gene. The variations you can have are C/C, C/T and T/T. This is what we consider to be the "raw data," which requires interpretation. If you only have your raw data, the table shows what that translates to:

rs5400 (C/C)	normal	typical taste for sweet
rs5400(C/T)	heterozygous (one copy)	increased taste for sweet
rs5400(T/T)	homozygous (two copies)	increased taste for sweet

More than a decade ago, this same gene variant was shown to be associated with increased risk of diabetes type 2. Back then, researchers did not fully understand this connection but since then, scientists have determined that the receptor this gene codes for acts as a glucose sensor (GLUT2 to be exact), controlling the response to sweet.

In 2008, researchers thought that glucose sensing in the brain was involved in regulating food intake and they tried to find out how the mechanism worked. They examined two distinct populations in Canada to determine if there were differences in young healthy adults, aged 20–29, or older adults with early stage type 2 diabetes not controlled with medication, aged 42–75 with an average BMI (body mass index) of about 30 (which categorizes them as obese).

The results of the study showed that in both healthy and obese diabetic patients, the subjects with the variations in the gene (the SNP) displayed higher intakes of sugar. This contradicted the previous view that biological mechanisms related to glucose regulation and insulin resistance were involved in sugar intake. This study concluded that higher sugar intake is more likely due to glucose sensing in the brain…thus finally identifying that elusive sweet tooth!

To my patients who have the sweet tooth SNP on their genomic profile, I try to offer sensible, easy-to-follow advice. I'll discuss more in part 3 but for now I will repeat what I always say to my family and friends. This is how you are "wired," so you should be public about your SNP. Own it, and say, "Yes, that's me. I have a sweet tooth" and tell people not to show their love for you with gifts of milk chocolate and jelly beans!

Satiety—The Bottomless Pit

✳ FTO-rs9939609

The FTO gene, sometimes dubbed the "fat gene" located on chromosome 16, has been the subject of numerous studies. In population after population, from Finnish to Mexican to Chinese, the FTO gene is consistently associated with obesity. This, of course, has led researchers to try to figure out why people who have this variant are more likely to be overweight. The issue has been whether the observed association is the result of a slower or impaired metabolism, or increased caloric intake.

In 2008, a group of researchers proposed a well-designed study to find out if it might be related to appetite control. Investigators enrolled more than 3,000 children in the United

Kingdom to better understand what they called "habitual appetitive behavior." The assessment was completed by using two scales, Satiety Responsiveness and Enjoyment of Food, part of the Child Eating Behavior Questionnaire. This assessment form has been validated against objective measures of food intake. The findings demonstrated that children with two copies of this gene variant, who were AA, or homozygous, experienced more difficulty feeling full than their counterparts. This study concluded that the variant was involved in appetite control. This study showed that the FTO variant did affect appetite control.

And, of course, this can be explained by genomics and now, we have even more complex data clarifying these variants. While the SNP rs9939609 is the most widely studied variant in the FTO gene, we know there are many others and that they are often inherited together. In 2015, another SNP rs1421085 was suggested to be the causal SNP for the FTO gene. The bulk of the research studies have been conducted on the rs9939609, but in the near future, there will be more research on all these related SNPs. I will discuss the FTO gene in more depth in chapter 5, How your genes affect your weight.

If you have had genetic testing to confirm that you have this SNP or you just never feel full, you need to find ways to help you feel fuller after eating a meal. Again, I'll discuss this in detail in part 3 but one smart tactic is to consume foods that are likely to make you feel fuller. For example, when you eat an apple before dinner, you're more likely to feel full because the fiber in the apple (when accompanied by water) expands in your stomach.

Snacking—Are You a Grazer?

✳ LEPR-rs2025804

The pattern of our eating is a key factor in whether we're able to remain within a healthy weight range. For obvious reasons, if we are continuously snacking, or consuming high calorie foods, then we end up gaining weight and eventually can become obese. Clinicians and researchers have long observed these dietary patterns in some obese people. In 2007 a group of researchers published a study in the medical journal *Diabetes* that examined which gene variants were present in people who had quite extreme eating patterns.

Knowing the hormones involved in controlling frequency and quantity of food intake, the study authors were able to design an effective experiment to uncover the genetic factors at play. Cholecystokinin (CCK) and leptin are the two hormones that are known to regulate these factors. CCK can be thought of as the satiety regulator, while leptin is involved in eating behavior and is considered to be long-acting in energy metabolism and food intake. Leptin, when working properly, suppresses food intake, which thereby regulates weight management. Therefore, scientists expect that when variations occur in the gene for leptin or the receptor that leptin binds, they would observe people having difficulty maintaining a healthy weight.

This is exactly what was discovered in that 2007 publication, the European Prospective Study into Cancer and Nutrition (EPIC) study, which involved more than 17,000 women aged 49-70 from the Netherlands. These subjects had recorded detailed accounts of their diet and eating habits.

The researchers selected women who were categorized as consuming extreme meal size or frequency defined as within the top fifth percentile and were classified as having a BMI above 33kg/m^2, which would fit the category of obese. Genetic studies were done on 267 women and then compared to a randomly selected control group from the same study. The results showed that women with a single nucleotide polymorphism in the leptin receptor, rs2025804, were more likely to display the extreme snacking pattern.

The takeaway message is similar to what I've discussed about the other behavioral traits, such as eating disinhibition. The drive for snacking can be genetically predetermined. If you are someone who seems to be snacking constantly and has trouble controlling this type of eating behavior, then it could be in your DNA. What can you do about it? This is where the targeted approach is key. Telling people who have this SNP that they should only eat at meal times would be a mistake because that tactic doesn't match their genetic variant. When you are primed to be a grazer, it's particularly difficult to only eat three square meals per day. For sensible ways to manage this SNP, read part 3.

CHAPTER 4

#2 EXERCISE: Exercising Your Genes

Exercise has long been recognized as a key component to health and wellness. There is tremendous variation in the types of exercise programs available, and especially before the start of the summer beach season, when we're all in shorts or bathing suits, we'll hear about the latest and greatest program designed to solve all our body flaws. There are essentially four types of exercise: endurance, strength, flexibility and balance. Some fitness experts and programs emphasize one type more than others. However, what works for one person may not work so well for another. This is the unifying concept I will continue to emphasize in this book and in my practice. We are unique and cannot expect uniform outcomes. Just because your neighbor is getting killer results from her new weight-lifting program does not mean that you will be rocking that same set of toned abs. In addition, there's a reason your coworker's cholesterol profile improved dramatically on her program while yours barely budged. It's all about your genes.

Again, this is where genomics can be of huge benefit. Instead of using trial and error, changing fitness center memberships year after year, bouncing from one personal trainer to another or buying the latest infomercial product, you should take advantage of the emerging tools in the field of genetic testing. Although the science has been around for nearly two decades, it is now more widely available. Accept that the way your body responds to exercise is primed by your genetics. Don't be discouraged and give up hope; armed with this realization, you should follow your own set of rules. Research has shown that some people have genetic variants, or SNPs, that result in more fat loss in response to endurance exercise, while others can store more fat when they lift weights. I hope that this information is reassuring and may provide you a sense of relief. Understanding their genetic response to exercise has proved life-changing for some of my patients.

Another concept that is crucial to understand is that exercise is of vital importance, not simply because it conditions our cardiovascular system and helps us reduce our fat mass, but because it communicates with our DNA. Exercise is actually the equivalent of an environmental factor like diet, stress and exposure to toxic chemicals. It is part of the complex interplay of nature and nurture, genetics and epigenetics. By understanding our genetics from reading part of our book of life, and then optimizing the environment by practicing forms of exercise that are aligned with our genetics, we will be on our way to reaching our genetic potential. This chapter will discuss some of the better known exercise SNPs and emerging concepts in genomics.

Strength Training: Could Lifting Weights Make You Fat?

✳ INSIG2-rs756605

Perhaps surprisingly, the answer is possibly yes, but only for certain people. When I first heard about this research, I was stunned. How could exercise cause someone to gain more fat? I thought perhaps the research was somehow flawed. But when I did more digging, I found the science was convincing. And, I knew that I had more than my share of patients over the years who complained about this phenomenon. They would tell me that the more they worked out intensively with weights, the fatter they felt. They would say their legs and arms did not look well-defined and.in fact, appeared even heavier, exactly the opposite results they expected from starting a weight-lifting program. They were puzzled by this outcome and so was I…until I began using genetic testing in my practice.

Once I started incorporating a genetic profile for wellness into my assessments of patients having trouble with weight management, I learned about the gene INSIG2, also known as the Insulin Induced Gene 2. For people who have a genetic variant of the INSIG2 gene, or "SNP," when they include resistance training into their workout, there is an association with increased fat volume after 12 weeks of exercise.

Some of you may be surprised or even shocked, but my patients who have participated in a serious resistance program have experienced these "results" firsthand. The patients who have not noticed it to be an issue did not incorporate intensive weight training into their exercise routine or didn't have much of an exercise routine.

You may wonder what kind of a program was used in the study. The protocol was described this way: "Briefly it consisted of two 45–60-minute sessions per week for 12 weeks. Each session was supervised by an exercise physiologist professional or a trained student. Before each session, participants warmed-up with two sets of 12 repetitions of the biceps preacher curl and the seated overhead triceps extension. Each session included dumbbell biceps curls, dumbbell biceps preacher curls and incline dumbbell biceps curls, overhead dumbbell triceps extension and dumbbell triceps kickbacks. The amount of weight was aggressively increased during the 12 wks." That's a very typical resistance program and many people follow much more intense programs than this one.

Another very significant point in this research is that it only applies to men so far, and there is insufficient evidence to determine whether women are similarly affected. Speaking as a physician who specializes in precision medicine and is working with this type of testing daily with many female patients, I am fairly certain this applies to women as well. It is only a matter of time before we see research confirming it. Some of my female patients have been working out with heavier weights more frequently and even daily, and/or spending longer periods of time—more than two hours—at the gym. More studies are needed to evaluate the impact of this type of more intense resistance program, especially in women.

The genetic testing we employ in my practice also provides information on the benefits of endurance training. There are also gene profiles associated with enhanced response to endurance training. Again, I can't emphasize enough how beneficial it is to understand your genes. I truly believe that information is power and the more we know, the more we can guide our efforts in the direction of optimal health. Precision lifestyle medicine allows us to deliver true HEALTH care.

Endurance Training: Better Results for Some People

* LIPC-rs1800588

* LPL-rs328

* PPARD-rs2016520

Exercise that is categorized as endurance by definition increases heart rate and breathing rate, thereby conditioning the cardiovascular system. When we talk about endurance, we are talking about a program that is adjusted and changes over time according to the needs of the individual. As our body adapts to a certain intensity of exercise, it becomes conditioned to that intensity, and the program needs to be increasingly more difficult in order for the body to reach the target heart rate level. The opposite concept occurs when the body becomes "deconditioned" or "out of shape" from inactivity. For example, I often see patients who had been runners but then stopped exercising completely for years. When they first start running again, they will be unable to run at the same level of intensity which they ran before they stopped. That's why I advise people who haven't exercised in a while to ease into a program more slowly than they would like.

Over the years, scientists had observed tremendous variability in the amount of fat lost in response to endurance exercise and had even conducted studies in twins and families suggesting that there were indeed genetic differences that accounted for the different outcomes. In 2001, researchers conducted a landmark experiment in which they studied specific genetic variants in a group of participants already enrolled in the HERITAGE Family Study. This study included both white and black male and female subjects as well as their adult offspring to observe the effects of a 20-week endurance program on metabolism and cardiovascular function.

The endurance program consisted of three exercise sessions per week on a stationary computerized bike that controlled power output. During weeks one to three, they exercised for 30 minutes at a heart rate associated with 55% of their pre-training maximal oxygen (max O_2) consumption. Over the next 10 weeks, they increased to a heart rate of 75% of their max O_2 to 50 minutes. They maintained that level of exercise for the last six weeks of their 20-week endurance program.

The results demonstrated that there was, indeed, a difference in the amount of fat lost, which was tied to a specific genetic variant, or SNP, called LPL-rs328 or which is referred to as the LPL-S447X polymorphism. Across the board in men and women, there was significant reduction in body fat. As is typical with endurance exercise, as a group, the men lost more abdominal fat mass than the women (which is incredibly frustrating to my female patients). But interestingly, the findings related to the genetic variant only applied to the women and were much more significant in black women. White women who had the rs328 SNP had greater reductions in body mass index, total body fat and body fat percentage. Black women who expressed the rs328 SNP had a six-fold increase in the LPL activity (an enzyme related to fat burning) which was correlated with fat loss. These results help to explain, at least among women, why some of us get much better results than others with an endurance program.

Research has also shown that some other SNPs related to exercise can also influence improved health outcomes in response to an endurance-style program. One of these SNPs which has been studied more thoroughly is LIPC-rs1800588, which mediates improved cholesterol/lipid profiles in carriers who are on a consistent endurance exercise program. This SNP is relatively common in the general population and helps explain why, for most people, exercise can improve cholesterol levels. However, I have seen cases where the patients consistently follow an exercise program, yet it doesn't help improve their cholesterol levels. The exercise program could be beneficial in other ways but if these people had a genomic test prior to starting a treatment program for abnormal cholesterol, we would know not to rely on exercise as a primary therapeutic strategy. This SNP also helps explain why there has been confusing results in some research studies that looked at the role of exercise in treatment of abnormal lipids.

Another SNP, PPARD-rs2016520, has been shown to frequently appear in elite athletes, which suggests an association with improved performance. The genetic potential of an athlete tends to include at least 10–15 identified SNPs, including this one. I want to stress the word potential because we know the performance of athletes is, like most "conditions," the result of interplay of a person's genetics and his or her environment.

Aerobic Capacity

✳ PPARGC1A-rs8192678

This gene codes for a protein that is involved in many functions related to energy, as well as the metabolism of glucose and fatty acids. Researchers have known for some time now that specific SNPs in this gene can affect how a person responds to endurance exercise and also impacts glucose control and risk of diabetes. In other words, if your genetic profile shows that you have this SNP and you are currently obese, you are more predisposed to developing elevated glucose levels than someone who does not have this SNP. However, if you are not obese, there is an inverse relationship and you are less likely to develop high levels of blood sugar.

This is not simply another reason you should maintain a healthy weight. The information obtained from this research goes much further. It supports the theory that our environment influences our genes. The set of lifestyle choices that led to someone becoming obese created a condition that can "turn on" genes, which increases the likelihood the person will become diabetic. Furthermore, it also helps us understand why some people who are obese experience elevated blood glucose, while other obese people don't have high blood glucose levels. For many years we have known obesity is the single most important risk factor for the development of type 2 diabetes. Now genomics is helping us explain why. When one of my patients tests positive for this SNP, I know she has to be extra vigilant in following the care plan so she can maintain a healthy weight to greatly reduce her risk of becoming diabetic. It's hard for most people to commit to the necessary lifestyle choices it takes to sustain a healthy weight. However, we're also primed to avoid pain, so knowing that diabetes could be the outcome of poor eating habits can motivate us to choose an apple instead of a bag of pretzels for a snack.

Beyond the metabolic effects, this SNP also has relationships with aerobic capacity. We know from some older studies that this SNP is less commonly found in groups of elite endurance athletes because it is associated with reduced aerobic capacity and less of an increase in anaerobic threshold with endurance training. Consequently, people who have this SNP will find it harder to perform aerobic exercise compared to someone of the same age who exercises at the same level of difficulty. In addition, aerobic capacity declines with age because of decreased heart output, ability of oxygen to be carried around in the blood, the density of tiny blood vessels and changes in muscle mass and metabolism. If you have this SNP, you should continue a consistent exercise program. As you get older, you shouldn't stop cardiovascular activity because of its perceived difficulty. **Of course, if you are not already exercising or find that exercise is difficult, please make sure you check with your doctor before starting or continuing your program.

Muscle Power: The Sprinter Gene or Weightlifters' Gene

✳ ACTN3-rs1815739 (often referred to as R577X)

Variation within the ACTN-3 gene, which codes for a protein called α-actinin-3, may play a role in the ability of world-class athletes. A-actinin-3 is a z-disc structural protein found in fast glycolytic muscle fibers which are responsible for the forceful and quick muscle contraction that is necessary to excel in sports like sprinting and weight lifting (in which are the groups of athletes that have been the focus of most of the research studies). One-quarter of the world's population has the genotype XX and produces no α-actinin-3.

A 2008 study involved 486 Russian power-oriented athletes (weight lifters) who met regional or national competitive standards and assessed the frequency of the different types of ACTN-3 polymorphisms. In a comparison of the power group to a control group, scientists found that 6.4% of the athletes had the XX genotype compared to the control group at 14.2%. Furthermore, only 3.4% of the most elite athletes had the XX genotype, leading the researchers to conclude that *ACTN3* R577X polymorphism was associated with power athlete status in Russians.

For the general population, having the XX genotype doesn't make much difference in our daily lives. In addition, this is only one of hundreds of genes that play a role in performance, which is also impacted by environmental factors like nutrition and lifestyle habits. However, when you are a top-level athletic competitor, these tiny variations can impact competitive results. This research is more evidence of the complex relationship between our genes and environment. Further study will provide information that can be used to enhance our daily life experiences.

Genetic Testing for Selecting Sports, Activities, and Training

Widespread use of this type of testing is still in its infancy, but has the potential to help guide training programs and possibly give direction in selecting sports and activities that may be aligned with genetic potential. What if you could identify early on whether your child had the sprinter's gene, or genes associated with enhanced endurance, agility or aerobic capacity? Getting this data is now in the realm of medical possibility rather than being science fiction.

Having been actively involved in youth sports with our own kids, some time ago my husband asked me whether it would be possible to develop a profile that could help guide families to select sports aligned with their children's natural talents. Of course, genetics is thought to be only about 30% of the equation and this type of testing shouldn't be used to deter children from pursuing sports that interest them. However, wouldn't it be nice if you knew early on that if your child was interested in skating you could narrow down genetically what sport—speed skating, hockey or figure skating—was the best match? You might even focus on a particular position within the sport.

The utilization of genomic testing in elite athletes has already begun, with top athletes seeking out this type of wellness testing to help develop their individual training programs. Olympic level athletes and professional soccer players around the world are already making headlines about using genomic profiling to gain a competitive edge with their nutrition and exercise programs. This is a smart strategy for athletes and their trainers. If, for example, an athlete had a SNP which caused impairment in methylation of folate (commonly known has the MTHFR SNP), then it would be very helpful for that athlete to take supplemental activated folate- L-5-MTHF to support his metabolism.

I recently spoke with an athletic director for a large state university who agreed that programs could greatly benefit from utilizing genetic testing rather than relying on a uniform training program used for all athletes. Knowing precisely how one person responds to endurance training and weight training would allow for customization of a training program to meet his or her unique requirements. I believe genetic testing will become commonplace for elite athletes and in professional sports in the not too distant future.

CHAPTER 5

How Your DNA Influences Your Weight and Metabolism

W hat you choose to fuel your body with every day is ulti- mately the biggest decision you can make regarding your health status. With some exceptions, most of us who live in the industrialized world have the luxury of picking what foods we are going to consume on a regular basis; this is one of the foundational keys to creating long-lasting health. However, patients often complain that, despite having a generally nutritious diet, they feel poorly, struggle with health problems or have difficulty maintaining their ideal weight. Usually, because I've had so much experience coun- seling patients on nutrition, I can figure out what's wrong with their dietary habits.

More recently, with the advent of genomics, I have incor- porated genetic testing into the assessment. Using a simple salivary test, I can tell a patient whether there are any inborn issues with his metabolism and what ratio of carbohydrates, protein and fat best matches his individual genetics. This allows physicians to create a designer genetic diet.

Shifting the ratio of the macronutrients allows us to eat many of the foods we are accustomed to eating by modifying the quantity we consume. And, if you have SNPs that lower the rate at which you burn calories, then eating a slightly lower calorie diet and/or increasing your level of exercise can be advantageous.

Of course, many patients come to me with chronic conditions such as fatty liver disease and diabetes. Utilizing diets based solely on genetic testing does not take into account these types of metabolic conditions. Physicians can evaluate both genetic and metabolic tests to recommend proven nutritional lifestyle strategies for their patients.

In addition, further evidence has shown the impact of diet on the bacteria in the intestines (the microbiota) and its ultimate impact on the body. A healthy eating plan will address all of these concerns. Furthermore, it may turn out that the interplay of the foods with our genes will involve critical signaling from these bacteria. Research studies on the microbiome—a term used to refer to the microbial organisms that live in and on our bodies—is almost at pace with human genomics, and we are at the brink of major advances in this field.

In the next two chapters, I will discuss what we know about genetics, your metabolism and obesity. Once you get the basics, then comes the good stuff—dietary profiling, genetic matching diets and how your body reacts to certain foods based on your genetics. This scientific field is advancing rapidly and every year, new SNPs are discovered, enhancing genomics panels, thereby adding new pages to our "book of life." The final chapter in this part will explain the latest research on the human microbiome that can also be assessed using the advances in DNA technology that have allowed us to learn so much about our human genome.

How Your Genes Can Affect Your Weight

Testing for genetic markers associated with obesity provides a very valuable piece of the puzzle. When I counsel patients who are struggling with weight issues, I consider these tests critical in understanding the challenges people face. Surprisingly, many of the patients who are categorized as obese turn out not to even have these genes.

I remember the case of a 52-year-old woman who worked at one of our local hospitals for whom I ordered a genomic profile. She came to me for a consultation because she had been dealing with weight problems for most of her adult life. When she returned to the clinic so I could discuss the results of her profile, she was taken aback. Not only did she not have genes associated with obesity, she also had a fast metabolism, no issues with adiponectin levels (a hormone related to fat loss, which I will discuss on page 49) and only one of the eating behavior SNPs, eating disinhibition. However, she did have the weight loss–regain variant. Once we discussed her relationship with food in more detail, the test results made sense to me. During periods of stress and when she was feeling good (as on her recent summer vacation), the patient would eat larger portions of food, including bread and pastries, which were not a good match for her genetics. Consequently, during these times when she went off her "diet," she felt she gained weight rapidly. She always assumed there must be something wrong with her metabolism and that she was destined to be overweight like other members of her family. However, once she understood the true nature of the problems she was facing based on her genomic profile, she was excited at the prospect of finding a way to combat her weight gain. First, switching to a nutritional plan that is appropriate for her genetics will help her reach a healthier weight. She is a match for a Mediterranean diet and sourcing her fats should come mostly from monounsaturated fats. Next, because her body does well with consistency, radical increases in the number of calories she consumes will cause her to gain weight. She needs to follow a plan where she eats enough of the right foods daily so that she doesn't feel deprived or on a "diet" so she will stick with the plan over the long term. Third, she needs to have a behavioral plan in place to deal with her tendency to display eating disinhibition. She knew there was a cause for her difficulty maintaining a healthy weight and she could take steps to change. She was empowered and stated emphatically that she could do this and felt she now had control of her life. One of the great blessings of my work as a physician is to witness patients such as this one transform their health when they are armed with the right information.

THE OBESITY GENES

✳ FTO-rs9939609

In a landmark paper published in *Science* in 2007, the fat gene, or specifically the "fat mass and obesity associated" gene, was dubbed one of the most significant discoveries in years. A specific variant in the FTO gene that had been known to predispose people to diabetes by an associated increase in body mass index (which is a measurement that defines obesity) was shown to result in a 1.67-fold increased chance of someone becoming obese. Being homozygous (if you remember from our genetics overview), or having two copies of this SNP (genotype here is AA) signals the highest risk someone will become obese. This SNP is specifically associated with an increased fat mass and can be seen in people as young as seven years old.

Of course, people have suggested for years that obesity runs in their family; I will discuss genetic factors now and epigenetic factors in part 2. This study finally provided evidence of how obesity can be genetic. Further studies have shown that epigenetic changes, marked by patterns of methylation, are involved in expression of these genes. This distinction is key: A person may carry this variant but it may not get turned on. This is the beauty of our genomes. We can carry these variations but our environment determines whether the genotype ever gets expressed. I have seen many patients with genes for obesity who are not obese. And I have seen many patients who do not have genes associated with obesity who are obese.

Hundreds of studies on this genetic variant have been conducted since 2007. We have also seen some other variants on the FTO gene that also appear to increase the risk of obesity and in turn, diabetes. There have been numerous population-specific studies and this variant has been observed in people of European descent as well as Chinese, Japanese,

Korean, Indian, African-American, Mexican, Brazilian and even Roma/gypsy populations. This genetic variant is also involved in other health issues, including miscarriage and polycystic ovarian syndrome (PCOS).

I want to emphasize again that even if you have this variant, you are not destined to be obese or struggle with weight all your life. And taking it a step further, if you know you have the "fat gene," science tells us you have an advantage. People who have had genetic testing and find out they carry this type of risk turn out to be healthier than people who have not had this type of testing. How is this possible? It can be explained by the choices you make every day. Research shows that you will be more likely to exercise, make better dietary choices and overall live a healthier, cleaner lifestyle. Regardless of whether you have already delved into your own genomics, you can improve how you feel every day by practicing better self-care. It seems simple, but knowing that every choice you make influences the expression of your genes is an important concept that is often underplayed. Reducing your response to stress, eating more fiber and going for a brisk walk are all activities that impact your genes. And by having a genomic profile you can enhance your program by making your choices specific to your individual needs. That's when the game really changes!

✳ MC4R-rs17782313

The other main gene variant involved in obesity is the MC4R gene, which is short for Melanocortin 4 Receptor gene. Unlike the FTO gene, MC4R seems to act as more of a polygenic contributor to obesity, and requires interaction with other genes to result in obesity. When I am reviewing a patient's genomic profile and see that he or she is presenting with both FTO and MC4R SNPs, I know the patient is probably having trouble maintaining a healthy weight, even when the individual is doing everything right.

The evidence is not conclusive on whether these genes are associated with diabetes and it seems more likely that the diabetes risk is the result of elevated BMI, rather than the actual gene variants, since the risk goes away when the BMI is reduced. Again, as I explain to my patients, it really does matter what lifestyle choices you make. You can indeed influence your risk of disease by maintaining a healthy weight. Having these "fat genes" does not guarantee you will develop diabetes.

Furthermore, the good news is that these gene variants are dependent on very specific environmental factors, particularly dietary patterns. An excellent study published in 2012 examined the effect of a Mediterranean diet on the expression of FTO and MC4R SNPs and risk of type 2 diabetes in more than 7,000 subjects. The study found that people who had either of the gene variants and didn't follow a Mediterranean-style diet had much a higher rate of diabetes regardless of their BMI. But when people closely adhered to a Mediterranean diet and were "blessed" with the FTO and MC4R SNPs, then the risk of diabetes completely disappeared! Here is the conclusion the researchers stated in their paper:

*"These novel results suggest that the association of the FTO-rs9939609 and the MC4R-rs17782313 polymorphisms with type 2 diabetes depends on diet and that a **high adherence to the Med Diet counteracts the genetic predisposition.**"*

As a physician practicing **Precision Lifestyle Medicine**, this study provides definitive evidence that lifestyle matters, and that the expression of your genes is indeed dependent on the lifestyle choices you make. It's also an easy way to explain to my patients how lifestyle medicine is effective and prevention really is the best form of medicine!

There is tremendous complexity in how these gene variants interact and produce phenotypic effects. The focus of this book is on wellness and not disease states, but I would like to point out that there are many gene variants that can be utilized for testing various types of cancer risks, like the familiar BRCA genes that women with a family history of breast cancer are increasingly utilizing for risk assessment. Even if this test turns out negative, it does not mean these women have little risk of developing breast cancer. Many women who have breast cancer do not have the well-known BRCA genes.

Cancer is an area in which there is a tremendous amount of research focused on gene therapy. Cancer is extremely complex and there are many identified associations of obesity with various types of cancer. For example, we know that there is a link between obesity and risk of breast cancer, especially in post-menopausal women. A study published in 2013 demonstrated the involvement of both the FTO and MC4R SNPs in increasing the risk of developing breast cancer more than four-fold. Interestingly, on its own, the FTO SNPs did not increase risk, but the MC4R SNP did. Much more research needs to be done on this specific SNP.

How Your Genes Can Affect Your Metabolism and Digestion

IS YOUR METABOLISM FAST OR AVERAGE?

✽ LEPR-rs8179183

Most of us think we have a pretty good sense of whether we have a fast or slow metabolism. However, in my experience, both personally and in working with patients, I have found that what most people think about their metabolism is not what their genomic profile reveals. Metabolism is actually reported as "fast" or "normal," which we would probably all agree unfortunately means slow. So yes, your metabolism is also genetically predetermined. Of course, some people may have other metabolic issues like an under-functioning thyroid gland that can contribute to their body's seemingly slow metabolism, (but in my clinical experience, I would estimate that the maximum amount of weight attributable to an untreated thyroid patient is less than 20 pounds).

The SNP that affects your metabolism involves the leptin receptor. Leptin is actually a hormone that exerts its effects through a rather complex mechanism that affects fat burning in the tissues and metabolic rate controlled in an area of your brain. Research conducted in the Quebec Family Study examined a number of gene variants in the leptin receptor. It turned out that individuals who had two copies (homozygous) of this SNP exhibited a higher metabolic rate.

My husband and I had been together for close to 20 years, when we had our genetic profiles done. For that entire time, we had a running commentary about my "fast metabolism" and his "slow metabolism" because it often seemed that I could eat as much as him and still maintain a healthy weight. Boy, were we wrong about this SNP! It turned out that he had the "fast" metabolism and I was the one with the "slow" metabolism. And in fact, it appears that the overall dietary pattern that I followed was, for the most part, a genetic match for me while his diet was not so great a match.

Knowing your metabolism is of great value with respect to your mindset and relationship with food. Being cognizant of whether you have a fast or average metabolism is yet another piece of information that gives you power to live according to your code. If you have a fast metabolism, you then know that your body can "handle a few more calories" than would be expected of someone of your size and activity level. As a result, you may not stress as much about your food intake. And the more you can reduce stress and embrace an attitude of gratitude you'll be doing your genes a favor and allowing them to express your true health potential.

On the other hand, if you have an average metabolism and know you are not rapidly burning the 'fuel' you are consuming, you must incorporate this mindset into your relationship with food. You should adopt the attitude that your body is "conservative" with its fuel. You savor the food you eat, chew more slowly, taste the full essence of your meal and know that the process of eating is another opportunity to reduce stress in your life. There was an interesting study about attitudes toward food reported in the journal *Explore*. Researchers examined seven factors involved in eating and determined that certain eating styles were related to obesity and overeating. Many people just focus on what they eat, but patterns and attitudes toward eating may be a contributing factor in whether you're able to stay in a healthy weight range.

Researchers developed questionnaires to measure food, nutrition, and eating themes in more than 5,000 participants. The results helped the researchers identify seven eating styles that were independently related to overeating, and five that were significantly associated with people being overweight or obese. By analyzing the results of the questionnaires, researchers developed an integrative eating score, where those at the lowest end were the most obese. The eating styles of this group included: (1) emotional eating; (2) consumption of more processed, fast, sweet, and fried foods and less fresh, whole grains, fruits and vegetables; (3) paying less attention to sensory and spiritual aspects of eating; (4) focusing on self-judgment and feeling guilty about overeating (also called food fretting); (5) more likely to eat in a hectic, tense atmosphere; (6) more likely to eat while doing other things; and (7) more likely to eat by themselves.

This study shows that we should take the wisdom of other cultures and the way most of our society ate in the past: respect the world that provides us sustenance, express gratitude for nutritious, fresh food, appreciate the surroundings in which you eat, eat only when you are hungry, take time to prepare nutritious meals, and enjoy them in a relaxing atmosphere with people you love, focusing on your meal and no other tasks. Turn off the TV when you're eating!

THE WEIGHT LOSS–REGAIN GENE

✳ ADIPOQ-rs17300539

No doubt, you are one or you know someone who is a yo-yo dieter. You go on and off diets but you end up weighing more than before you started the diets. The list of options is almost endless, from Weight Watchers, Atkins, South Beach, master cleanse, paleo and on and on. Perhaps the cycle started when you found out about a reunion or some other upcoming event where you would see an ex-boyfriend or someone you hadn't seen in a long time and you wanted to impress the person by fitting into that favorite dress hanging in the back of your closet. Or, it's early April and you want to lose 10 to 20 pounds to look good in your swimsuit for a summer vacation, so you go on a crash diet. And, it works. You shed those pounds rapidly by eating very few calories and overdoing it on exercise. You are happy. You did it! You look great in that dress; you are on a mini-high at that reunion. But, that night at the reception, you drink a few more glasses of wine than you are accustomed to and start eating more than you have been eating over the last six weeks. Then the next day, you order pancakes and bacon for breakfast, which leads to a day of consuming more and more calories. This trend keeps up and suddenly three weeks later, you have gained back those 10 pounds, plus two more!! You can't believe it. How could you weigh more than you did back in April? To make matters worse, this has happened during each of the last three years. You feel defeated and start to wonder why you even bother trying to get healthy!

If this sounds familiar, you may have a genetic variant that makes it easier for you to regain weight after you have been on a lower-calorie diet. Yes, there is a SNP for this behavior too! And if you have it, you are not a good candidate for a very low-calorie diet. These types of diets do produce weight loss but are not sustainable and you will end up gaining the weight back quickly, and then some.

I recently ran a genomic profile on an obese patient to help devise a precision lifestyle plan for her. She had none of the obesity genes and only one out of six of the eating behavior variants, but she did test positive for the weight loss–regain variant. I asked whether she had ever experienced rapid weight gain after going off a low-calorie diet, and, of course, she always did. She assumed there was something wrong with her metabolism.

She went on to tell me that in recent years she had followed a structured weight loss program that cycled through a four-week, very low-calorie plan phase, followed by a two-week phase in which she was supposed to eat significantly more calories, and two days before starting the low-cal cycle, she was advised to "load up" on carbs and calories to jump-start the next phase because of the contrast in calories. She said she would always end up gaining more in the two weeks than she lost during the four-week low-calorie phase. It was obvious to me that this plan was the exact opposite of what she should be following.

This is a striking example of how this type of dieting can lead to excessive weight gain. Severe calorie restriction when you have this SNP will always backfire. It is so important to consume a similar amount of food day to day, so your body knows what to expect on a regular basis. The key here is consistency.

Another concern I have for patients who have this SNP is when they follow a trend—the cheat day that some fitness and nutrition experts recommend. This plan encourages you to eat healthy six days a week and then eat whatever you want on the seventh day. I really don't think this is a sensible approach for anyone, but it's especially flawed for people who are "blessed" with the weight loss–regain SNP. This type of diet just sets them up for regaining the weight they lost during the week. Again, the body doesn't respond well to dramatic differences in calorie consumption from one day to the next.

This genetic variant may also be applicable in gastric bypass, whereby patients are on caloric restriction following surgery, but over time start consuming significantly more calories. The variant may explain why some patients who have had bypass have less sustainable results. For these reasons, genomic testing is extremely valuable in helping me to manage particularly challenging cases related to weight management.

The blanket approach to diet and nutrition therapy is outdated. Following the latest advice from the Internet without knowing whether it is a good fit for your individual needs is a roll of the dice. The tools available to us through genomics now explain why a "one diet fits all" approach never works. Beyond the actual dietary and exercise elements, eating behaviors must also be addressed, and again, this is highly individualized. What I know works for me may certainly not work for you. The more I utilize genomic profiling in my practice, the more I am convinced that it will lead us to tremendous advances in our ability to promote health and wellness.

ADIPONECTIN

✳ ADIPOQ rs17366568

Adiponectin is a hormone produced by fat cells that affects your metabolism in a very positive way. The effect of adiponectin in the brain is an increased metabolic rate without impacting appetite or food intake. We can also test for adiponectin levels with a simple blood test. Low levels are associated with difficulty with weight loss. But administering adiponectin is not a cure for obesity. It turns out that we have yet another gene variant that can affect your metabolism. Unfortunately, if you have even one copy of this SNP, it increases your risk of obesity. Research has shown that this SNP occurs at a rate of 11% in obese subjects but in only 5% of non-obese subjects.

To me, this data is encouraging and lends more support to the running theme of this book. While you can have genetic variants that increase your risk of obesity, that doesn't mean you are destined to be obese and there is no hope. This SNP does occur in people who are not obese, just at a lower rate, but that alone is good news. Just like the obesity genes, I have patients who have this SNP who are not obese.

Maintaining a healthy weight is the outcome of a positive interaction with your environment when your lifestyle matches your genes. For some people, this interaction comes easily with little conscious effort. For others, it requires herculean effort. I believe that combining genomics with proven lifestyle strategies is the most targeted and "easy" way to get it right with the least amount of effort!

CHAPTER 6

#3 DIET: Genetics and Dietary Profiling

Let me start by telling you what you won't read here. I will not suggest a "one diet fits all" plan and, by now I hope you realize that this approach is fatally flawed. Since I believe that genomics reveals what your body needs, it makes no sense for me to say, for example, that paleo is the ideal diet for everyone simply because it was the diet of our ancestors. In fact, the macronutrient profile of the paleo diet which (depending on whose advice you follow) contains more than 50% fat is, unfortunately, not a good genetic match for any of us. I don't want to waste any of your time bashing other diets. (As an aside, what I do like about the popularity of paleo diets is that they offer choices for people who have problems with dairy and wheat, and they promote using whole foods instead of highly processed gluten-free foods.)

The goal of this chapter is to review the science behind the various types of diets that are considered a genetic match based on testing variants within the human genome. Criticism that genetically matched, or "personalized diets," are also a fad has been put to rest by the results of a clinical study at Stanford University which concluded in 2010. The findings demonstrated that people following their genetic appropriate diet lost *two-and-one-half times as much weight* as people on diets that were considered genotype inappropriate. This is a very significant difference in weight loss. There were also secondary outcomes showing improvements in cholesterol levels. The people who followed the genetic matching diet got results that were visible but were not so apparent to the subjects in the study until they learned their blood test results. But at least equally important, they also improved their metabolic health, which is a key point that I would like to emphasize. I have seen many patients who have been on radical diets that resulted in weight loss but ended up impairing other metabolic markers of health status.

There is an entire field of scientific research called *nutrigenetics* devoted to the study of the role of genetic variation among individuals with regard to the very complex interactions between diet and health for the purpose of developing personalized dietary plans, also known as genetic matching diets. There is also another area of focus called *nutrigenomics* which involves examining how common dietary ingredients affect the genome in a more general way in terms of whole body metabolism. Here, scientists develop genomic techniques to help figure out what specific dietary elements do to our genes. The two fields are closely related; the more information we gather, the better outcomes we will see in the field of human nutrition with respect to the development of personalized diets and improvements in long-term health.

You should understand the concept that you are a genetic match to a certain balance of macronutrients in your diet and eating the right foods for your specific makeup can influence your overall health. I would consider all these dietary profiles healthy, as long as the macronutrients (fat, protein and carbohydrate) are derived from a whole foods diet. The standard American diet (SAD), which relies heavily on processed foods and limited to no fresh, whole foods, could be set up to technically match the macronutrient profile, but I am certain the long-term health consequences of eating a SAD diet that is a genomic match would negate any benefit achieved by matching up fat, protein and carb ratios. This is where epigenetics, metabolic and gut health come into play. We have seen time and time again the detrimental effects of the multitude of chemicals found in these foods, as well as the lack of health-promoting phytonutrients.

Now, I would like you to take a moment to think about the composition of your current daily diet. Are you following a very low-carb diet? Is that working for you? Do you feel good eating that way? Or, are you following a very low-fat diet? How is that going for you? Perhaps you don't adhere to any particular plan. But when you consider how you truly feel and you realize that you should feel better than you do, there is no better way to start than what you are doing multiple times a day—eating. The choices that you make, in many cases, every couple of hours, can make a dramatic difference. I can't even begin to tell you how many times I have seen patients transform their health by changing what they choose to fuel their body with every day. This is why food is the very foundation of health and vitality.

The foods we consume contain much more than just macronutrients—fat, protein and carbohydrate. Much like our genetic variability determines how we respond to these larger food components, we can also have different responses to other elements in our diet, like caffeine, alcohol, lactose (milk sugar) even the way we perceive the taste of food as bitter or sweet! This is all explained by the unique makeup of our DNA. Perhaps you hate the taste of broccoli and other cabbage family foods and can't understand how someone else could actually crave kale! Well, that's the result of a specific polymorphism for a gene that codes for bitter taste. If you have this SNP, you might also tend to add more salt to bitter foods to try to mask the taste. Sound familiar? If this doesn't sound familiar, I'm sure you know someone who feels this way about anything leafy *and* green.

Now, I'll provide an overview of the matching dietary plans, followed by some of the better understood SNPs that affect your response to other food elements. You may have already realized (like I did) that if you eat a certain way consistently for a long enough period of time, you just feel better overall. Or maybe the opposite has occurred; your friend tells you she hasn't felt better since she started her new eating plan but then you try it and feel terrible. If you can't seem to figure out what works for you, it may be because the eating styles you have tried are not a good genetic match. Check out these and make some mental notes...you might be able to identify why you have faced so many challenges. Of course, you can always test and not guess!

The Four Genetic Matching Diets

1. THE MEDITERRANEAN DIET

Macronutrient Breakdown
- Fat: 35%
- Protein: 20%
- Carbohydrate: 45%

This is the dietary profile that has received the most scientific examination from empirical evidence to controlled studies. The Mediterranean diet has been touted as one of the healthiest you can follow and it has been studied to determine its impact on a wide variety of health conditions. Traditionally viewed as a whole foods diet—rich in plant-based foods, vegetables and legumes, healthy oils and fats, especially olive oil, avocado and nuts and seeds, and protein primarily sourced from the sea and small amounts of dairy—this is also appealing in a culinary sense. Balanced meals in a Mediterranean profile are rich in flavor, not heavy, and relatively easy to prepare. This is the dietary plan with the most fat,, which is key to understand since some people require the healthy types of fats in this plan to function optimally. It's also worth noting that the amount of fat in the Mediterranean diet is about 2/3 the amount of fat in a typical paleo plan.

This is the diet for which my husband and I are a genetic match, which is great news for us because we love this way of eating and have been following this approach for many years. On occasion, we veered away but we have always felt better when we stick to the Mediterranean diet. For brief periods, I have also followed a vegetarian diet, a vegan diet, and a lower carb diet, but never felt as good as when I ate the Mediterranean way. Now that I know this diet is the perfect match for me, I really do stick to this general plan on a daily basis. Again, I would emphasize that my patients also say they have been able to follow these guidelines more effectively than any other diets they followed in the past. This supports the evidence we have seen with genomic testing and outcomes: people who have had genomic wellness profiles generally, as a group, make healthier lifestyle choices than those who have not.

2. THE LOW-CARB DIET

Macronutrient Breakdown

▶ Fat: 30%
▶ Protein: 30%
▶ Carbohydrate: 40%

Low-carb diets have been very popular in the dieting world for some time, especially if you include paleo diets in this category, and for good reason. Obesity rates in the developed world have soared due to the overindulgence in excess sugar and carbohydrates. An editorial published in the *British Journal of Sports Medicine* in 2015 concluded that the obesity epidemic has been caused by consuming too many carbohydrates, not a sedentary lifestyle, as some have argued. In fact, the authors titled the article, "It is time to bust the myth of physical inactivity and obesity. You cannot outrun a bad diet." But, the question remains exactly how much carbohydrate is too much? And furthermore, which carbohydrates sources are really harmful?

The U.S. Department of Agriculture and the Department of Health and Human Services issued new guidelines in early 2016, advising people to limit their intake of added sugar to no more than 10% of their daily calories. Most of the added sugar consumed by Americans comes from sweetened beverages, like soda and sports drinks. But, refined carbohydrates in the form of flour, which is used in the production of the multitude of processed foods consumed regularly in the standard American diet, is a huge problem as well. When I define what constitutes healthy sources of carbohydrates to people, I suggest that most of their complex carbohydrates come from colorful vegetables (like leafy greens, bell peppers, and cucumber), then legumes, less sugary fruits like berries and green apples and limited starchy root vegetables and whole grains (like rice and quinoa).

I also refer to this as the 30:30:40 diet. We have successfully used this dietary profile along with medical food, which is a nutritional shake containing specific targeted ingredients approved by the FDA, for treatment of a medical condition such as impaired blood sugar control or inflammatory bowel disease. At my practice, we spend a lot of time explaining how different sources of carbs can have wide variation on the impact on blood sugar, referred to as the glycemic load of a food. Highly processed carbohydrates like flour have a high glycemic load, while non-starchy vegetables like broccoli have a low glycemic load. There are also other factors, such as combining a food with fat or protein, that affect glycemic load.

These days, I take another factor into consideration when prescribing dietary plans for my patients, and of course, you guessed it—their genomic profile. Some people can tolerate more carbohydrates than others. The people who match up to the low-carb dietary plan have better downstream metabolic responses when their carbs don't exceed 40% of their diet. I also advise these patients to make sure they are combining their carbs with protein and fat as best they can to optimize their glycemic response. Like any eating plan, once you get the hang of it, it becomes easy to follow the guidelines.

3. THE LOW-FAT DIET

Macronutrient Breakdown
- ▸ Fat: 20–25%
- ▸ Protein: 20–25%
- ▸ Carbohydrate: 50–55%

Eating a low-fat diet was all the rage before the low-carb craze hit. It was the most promoted diet everywhere for nearly 50 years. It developed from research in the 1940s showing that diets high in fat were also associated with high blood cholesterol, resulting in the medical community advocating a low-fat diet to prevent heart disease (although there wasn't any evidence that showed that this type of diet could have that effect). Then, on a much broader scale, Americans became encouraged to follow a low-fat diet for weight loss purposes, with everyone from government agencies to the medical community and the food industry getting on board.

In the 1980s, the food industry dominated the market with newer low-fat and fat-free foods, which were touted almost as free all together. The general public thought that if they ate a cookie, for example, (even if it were a stack of cookies) if it was fat-free, then it was just free, period.

Unfortunately, Americans were not getting thinner. In fact, the opposite was happening: obesity rates and, in turn, the rates of heart disease, were soaring. Low-fat diets did not hold the promise of improved health and vitality. The problem was not the actual low-fat content of the diet itself, but that this change in dietary balance resulted in elevated intake of refined sugars coming mostly from the consumption of sweetened beverages and processed foods. When food manufacturers removed the fat from that cookie, for instance, they increased the amount of sugar to enhance the flavor profile so that people would actually like the taste of that cookie. Furthermore, because of the low-fat mindset, most people thought they could eat as many of those cookies as they wanted because they were only counting grams of fat; they didn't consider the grams of sugar, or total carbohydrates, or for that matter, calories!

We are more knowledgeable about the role of elevated sugars in the diet, as well as fat, and more specifically, the types of fats in the diet and their role in health and wellness. We now recognize the detrimental effects of partially hydrogenated oils, or trans fats, and most people know how to avoid those manipulated fats that are added to processed foods to extend their shelf life. We also know that the "healthy" fats are important for our metabolism and there have been efforts to recommend people follow a diet higher in the healthier, unsaturated fats such as those found in olive oil and fish. But there is still no consensus on exactly how much fat is ideal for maintaining improved overall health.

It appears that the answer lies in our genetics and our overall metabolism. Some people do better on lower fat diets, in terms of their health outcomes. (We can even further define where those fats should be sourced from–super fats–which I'll discuss later in this chapter.) Lower fat here is considered in the range of 20–25% of total calories. But again, it does matter what the sources of *all* of those calories are. There's a vast difference in someone's metabolism if she consumes all her calories from fast food versus a diet of healthier natural foods. Fat still matters, more so to some people. For example, a study showed that some people gain more weight on higher saturated fat dairy products than low-fat dairy, and that effect is associated with a specific genetic variant. As a physician, knowing what I know about genetic variation and the impact of fat on the microbiome (more in part 2), I'm very concerned when I see health gurus touting very high-fat diets, upwards of 50-55% fat. Again, I can't emphasize enough that a one size diet does not fit everyone. If you have tried a diet and didn't get the results you expected, there's most likely a genetic and/or metabolic explanation for why the particular plan didn't work for you.

4. THE BALANCED DIET

Macronutrient Breakdown

- ▸ Fat: 25%
- ▸ Protein: 20%
- ▸ Carbohydrate: 55%

This plan is the highest in carbohydrates of all the programs outlined here and my patients who follow up are often shocked that they can consume this many carbs. We have become so accustomed in our society to labeling all carbs as bad that people often can't imagine that consuming carbohydrates would make them feel better. Again, keep in mind that the source of your calories really matters. This diet is essentially a variation on the low-fat dietary plan. Usually, people are consuming too much animal protein and not nearly enough vegetables and plant-based foods. Remember, all of those "superfoods" you routinely hear about are plants, not meats. Furthermore, we are now learning even more about how those indigestible components in plants help the bacteria in our gut thrive. Higher animal protein diets are actually associated with increased levels of potentially pathogenic bacteria in the intestinal tract.

Be aware, though, that this plan is a full 15% higher in carbohydrates than the low-carb dietary plan. Therefore, without genetic testing to determine if this plan is right for you, it may not be the best match for you. Don't compare results you get from a particular plan to a coworker's results from the same diet. One of you could be a genetic match and the other not. I see this every day in my practice. Patients come to me and say my neighbor followed "x program" and lost 30 pounds so they tried it but only lost five pounds. They want to know what's wrong with them. Of course, I routinely run a metabolic workup to screen for problems with blood sugar control and hormonal issues like low thyroid function. But I also utilize genomic testing, which will help determine the best dietary ratios of carbs, fat and protein, as well as data on the potential issues with metabolism like adiponectin and metabolic rate. Most experts in this field would say that blending this data together to devise the best lifestyle program for a patient is very much both a science and an art.

The Genetic Variants at Work in the Matching Diet

At the core of how I approach patient care and the relationships that I maintain with my patients is the philosophy of "doctor as teacher." I have always found that when people understand *why* I am advising a specific course of treatment, they are more likely to make those changes. That's ultimately what really matters. I believe that if you truly understand why a particular diet program is recommended by your doctor, you're far more likely to follow it. You'll be more committed to it than if you were simply handed a program without any explanation. To help you grasp how a genetic matching diet is devised, I will provide an overview of some of the SNPs on which the dietary plans are formulated. These are the variants that influence how your body responds to fats, carbohydrates and protein.

✳ ADIPOQ rs17300539

This weight loss–regain gene variant that was introduced in the last chapter is also of importance in determining the genetic matching diet. People who have this SNP (like me) have been shown to have a lower BMI as well as a reduced risk of obesity when they consume more than 13% of their calories from monounsaturated fatty acids (MUFAs). These are found in foods including olive oil, avocados and some nuts, particularly almonds, pistachios and peanuts, as well as sesame, pumpkin and sunflower seeds. The results held true across men and women in the study of more than 1,000 people.

✳ APOA2 rs5082

A genetic variant not covered so far, this SNP in the Apolipoprotein A2 gene has been the subject of recent study. In two separate groups of men and women, totaling more than 2,000, subjects were evaluated for their response to dairy in their diet as total dairy, low fat (<1%) or higher fat (>1%), to assess the different levels of saturated fats (which is the unhealthy fat found in animal sources). As expected from previous research, people with this specific gene variant had a higher BMI in response to the higher fat dairy.

This SNP is well known from an earlier 2009 study in which researchers demonstrated for the first time the interaction between genes and diet and their influence on obesity and BMI in three independent populations. That study showed that saturated fat was also the culprit for people with the homozygous CC genotype of the rs5082 SNP, which significantly increased their body weight, compared to the others in the groups.

✳ FTO rs9939609

This one may also look familiar to you—it's the same "fat gene" that I discussed in the last chapter. You may remember that I mentioned there was a study showing that people carrying this specific gene variant have an elevated risk of developing type 2 diabetes regardless of whether they were overweight. However, as long as they adhered to a Mediterranean diet, their increased genetic risk completely disappeared. How's that for proof of what following a genetic matching diet can do for your health!

❋ LIPC rs1800588

Associations have been drawn among SNPs within the LIPC, or hepatic lipase gene, body mass intake, dietary fat levels and circulating levels of blood lipids, particularly HDL cholesterol (also known as the good type of cholesterol). It appears people with this SNP are more sensitive to consumption of higher levels of dietary fat. In a study conducted in a group of Inuit subjects, the role of this gene variant with regard to the risk of developing colorectal cancer was studied, as dietary fat intake is an established risk factor for this disease. The results showed that not only was the risk of developing colorectal cancer affected indirectly through lipid metabolism, but an even more direct risk was demonstrated in this population. Researchers demonstrated again that SNPs in the LIPC gene were linked to higher BMI with higher dietary fat intakes.

❋ MMAB rs2241201
❋ KCTD10 rs10850219

These SNPs influence the interaction among carbohydrates, HDL cholesterol and your genes. MMAB is methylmalonic aciduria (cobalamin deficiency) cbIB type gene. For people who have the GG genotype of this MMAB variant, when they eat high carbohydrate diets, their beneficial HDL tends to be lower. If they consume diets lower in carbs, then their

HDL goes up. KCTD10, short for the 'potassium channel tetramerization domain containing 10', a gene that also contributes to regulation of HDL levels. For this SNP, the same GG genotype was more susceptible to the effects of higher carb intake.

Diets rich in carbohydrates have long been established as a risk factor for not only low HDL levels but also the development of atherosclerosis, the buildup of plaque in the arteries. This is the result of elevated insulin levels from the development of insulin resistance when blood sugar levels rise in response to meals containing high amounts of carbohydrates. Too much insulin remaining in the bloodstream triggers a cascade of metabolic effects that lead to the laying down of plaque in the artery walls, which puts you at higher risk for a heart attack or a stroke. When low HDL turns up on a patient's lab report, I want to review these SNPs to see if they are more sensitive to the effects of carbohydrate intake. If so, then the patient must follow a dietary plan that is lower in carbs to help reduce the risk of heart disease.

✴ PPARG rs1801282

PPARG is the peroxisome proliferator-activated receptor gamma gene. It codes for a factor that operates within cells and influences blood sugar and fat balance. A genetic variant in PPRAG has been associated in some studies with obesity, but not in other studies. More recent evidence suggests that this specific PPARG SNP is involved in someone being overweight and having elevated total cholesterol, but in women only, according to a study conducted with a group of Taiwanese subjects. Older research also demonstrated that in a group of women, polyunsaturated fats were important for achieving a lower BMI.

Unsaturated Fats and Essential Fatty Acids (EFAs)

Now you know that you are a genetic match to a certain balance of protein, fat and carbohydrate in your diet based on assessment of quite a few variants in your genetic makeup. But, before figuring out what diet is appropriate, there's additional data you need to understand. You can figure out which foods some of the calories should come from, in this case two different types of unsaturated fats. There are gene variants that determine your response to unsaturated fats; in other words, which sources of fat act as a "superfood" for you. This does not mean that one is good and one is bad. You should view them as "neutral" and a "superfood." You should tip the balance of fats in your diet to your superfood. For me, this is monounsaturated fats, which are found primarily in olives, avocados and certain nuts. This does not mean I am going to stop eating fish or using flax seed and chia seed, but instead I will make sure I use olive oil daily on my salads, eat olives with dinner and avocado most days of the week.

You've probably heard that omega-3s and fish oil are beneficial for your health. The term "fatty acids" is used in science to describe the nature of the chemical structure of these necessary compounds that our bodies cannot make on their own. These compounds found in the fatty portion of food are incorporated into the outer layer of our cells, and are critical for cellular function and influence our immune systems, especially with regard to inflammation. Some of these fatty acids go through additional metabolism within the body to get to the necessary end product, while others, like those in fish, are pre-formed.

Compared to our ancestors, our diet has remained fairly consistent in the total amount of fat consumed, but the proportion of fat coming from healthy omega-3s has been greatly reduced, while the unhealthier saturated fats and arachidonic acid (sourced mostly from animal products) has increased. Scientists have correlated these dietary changes with the rising rates of chronic diseases. For many years, I have been advising fish oil supplementation to the majority of my patients and have seen tremendous outcomes—everything from improvements in joint pain, anxiety and asthma to patients even reporting their hair appears shinier! There is now more evidence that deficits in healthy essential fatty acids may also be the result of genetic variation. Here we go again—your genes in action! This helps to explain why some people appear to be eating a diet that is apparently adequate in fatty acids but are functioning as if they have an EFA deficiency. There is also a routine blood test that I order to assess levels of omega-3 fatty acids. Let's take a look at the most SNPs that have the most research looking at how they affect your levels of fatty acids.

MONOUNSATURATED FATTY ACIDS (MUFAS)

✳ ADIPOQ rs17300539

I hope you remember this SNP. Briefly, it's the weight loss–regain gene variant and it has also been shown to be associated with a lower BMI when you consume more than 13% of your calories from monounsaturated fats. The main sources of monounsaturated fats are olives, olive oil, avocado, certain nuts and nut butters (almonds, pistachios and peanuts) and seeds (sesame, pumpkin and sunflower).

POLYUNSATURATED FATTY ACIDS (PUFAS)

✷ PPARG rs1801282

A study showed that women carrying this genetic variant resulted in lower BMI when the balance of polyunsaturated fats in the diet was higher than saturated fat consumption. The main sources of polyunsaturated fats are fish, omega-3 eggs, evening primrose oil (an omega-6), flax seed (omega-3), chia seed (omega-3), and walnuts.

OMEGA-3 AND OMEGA-6 FATTY ACIDS

✷ FADS1 rs174547

These omega fats in the diet are considered polyunsaturated fatty acids. The standard American diet is very low in omega-3 fats compared to the way we ate in the past. In general, we tend to consume small quantities of foods like fish, nuts and seeds, opting for more processed foods that lack these essential oils. What makes matters worse is that some people have a genetic variant that impacts the conversion process required for making other necessary metabolites in the fatty acid pathway. When I see patients who have this variant, I recommend they consume even higher levels of omega-3s and the healthy omega-6s, which for the most part involves nutritional supplementation with evening primrose oil and high quality fish oil.

Responses to Other Components in Foods

TASTE PERCEPTION

The ability to taste certain characteristics in food is also reliant on your genetics, so you have your parents to thank if you find the taste of certain foods bordering on offensive. For many years, I found it challenging to treat patients who were overweight or obese who hated vegetables. The core of a healthy dietary plan definitely involves eating a wide variety of veggies. Understanding the genetics behind perception of taste and the ability to taste some of the more bitter components of foods found in vegetables explains the individual nature of nutritional counseling. Now, when I find these genetic variants in my patients who have come to me for weight management, I am better prepared to help them find ways to overcome their inherent preferences that may have helped contribute to their weight problems.

BITTER

✳ TAS2R38 rs713598

✳ TAS2R38 rs1726866

The flavor of food is highly dependent on both genetics and specific components found in foods. In plants, these chemicals are called phytochemicals. Bitterness can be attributed primarily to a phytochemical called phenylthiocarbamide and similar compounds. The taste perception for bitter is a protective mechanism to help us avoid eating foods that could also possibly contain toxic compounds. (In today's world, we don't typically spend time foraging and eating unknown wild berries and plants.) The flip side is that, of course, the foods that can be very good for you, such as the leafy green cruciferous vegetables, better known as the cabbage family—broccoli, kale, chard, spinach, Brussel sprouts—also contain bitter phytochemicals. Some people really enjoy eating these foods, while others find them hard to consume because of the bitterness they perceive.

There are about 30 genes involved in bitter taste. Some of my patients who have these more common SNPs admit that they really don't like the flavor of these foods, but just go ahead and power through them just to get the health benefits, while others just can't get past the taste. People who have these genetic variants will also tend to salt these foods more to mask the bitterness to a degree. Adding salt can help improve the taste but if you consume too much sodium, you can impact your blood pressure and the kidneys. Unfortunately, with these SNPs if you have one, it is more common to have some of the others. Also, having one copy (being heterozygous) of this SNP results in the altered taste perception because it is a dominant.

I'll provide advice on how to prepare certain foods to make them more palatable in part 3, but for now think about using them with oils or seasoning to help improve the taste.

SWEET

✳ TAS1R3 rs35744813

Just like the taste for bitter, genetics influences the perception of sweetness in foods. When you first put a spoonful of ice cream in your mouth, that initial burst of sweet is the result of how your taste buds sense the sucrose (table sugar) contained in the ice cream. Food manufacturers have long understood the power of sweetness. How sugar tastes to me may be very different from how it tastes to you. This specific SNP is found less frequently in people of European descent. There are also other SNPs that influence the taste for sweetness.

If you have this SNP on your genetic profile, you have a decreased perception for sweet, meaning that foods that may be overly sweet may not taste that sweet and consequently, you tend to eat more sweet foods. On the other hand, if you don't have this SNP, when you consume a food with added sugar, it could taste overly sweet to you and you may not eat so much.

Consumption of sugar is much more complex than just the taste perception of sweet. Many factors play a role, including some of the behavioral SNPs such as the "Sweet Tooth" and "Eating Disinhibition" SNPs discussed in chapter 3. Furthermore, sugar is highly addictive and when people consume it on a regular basis, they tend to get hooked on it, much like certain classes of drugs. Thankfully, the federal government finally released dietary guidelines in early 2016 advising Americans to limit their added sugar intake to less than 10% of their daily calories. While this is definitely a step in the right direction to help inform people about the negative health consequences of excess sugar consumption, much work still needs to be done in this area.

Interestingly, research has also shown that the taste for sweet influenced by the taste receptor TAS1R3 also influences the level of fat and sugar consumption in the diet and has even been linked to obesity. There has been research looking at these SNPs in children and whether they may play a role in development of obesity due to food preferences. Other SNPs in the taste receptors for sweet have also been associated with increased risk of developing dental caries. When I counsel patients with this SNP, I know they need to pay particular attention to sugar intake and behavioral techniques related to consumption of sweets.

CAFFEINE

✱ CYP1A2 rs762551

Are you one of those people who always knew that you had an issue with caffeine? If you have more than one cup of Joe, you get a bad case of the jitters, or worse, heart palpitations, while your coworker or spouse downs three ventis every day and is not even fazed by it. If you drink a caffeinated beverage after your daily "cut-off time," you have trouble falling asleep. Guess what? There's a SNP for that too! If you have this genetic variant, then you have some degree of impairment metabolizing caffeine in your liver. Unlike people who do not have this SNP, you have difficulty with what is referred to as "caffeine clearance."

There have been countless studies on coffee and caffeine. One study says it's bad for us while another study says more than six cups of coffee per day is good for us! Which one should you believe? Neither…the flaw is that these studies did not take into account genetic variation in metabolizing the caffeine component of these beverages. Caffeine is only one component of the many hundreds of compounds that can be isolated from coffee. But it is a very important compound that has a drug-like impact. For that reason, it is used in migraine medications and in weight loss prescriptions. You need to be aware that there can be serious risk of adverse effects when you consume too much caffeine. For people who are poor metabolizers of caffeine, those effects can be seen at a lower dosage level.

A study of subjects who were slow metabolizers of caffeine showed that those who drank more than four cups of coffee daily had a significant increase in the risk of having a heart attack. Most of my patients who have this SNP were unaware of this heightened risk and the dangers of consuming large amounts of coffee. These patients have told me they don't notice any heart palpitations or jitters so they would have no reason to think they are harming themselves. Now, armed with information about this SNP, they can make an informed choice to reduce their caffeine intake. This is another great example of how a very simple lifestyle change—drinking one or two cups of coffee instead of five—can make a huge impact on your life. Even my husband has this SNP and had no symptoms when he drank more than four cups of coffee a day. After having his genomic profiling, he has curtailed his caffeine intake. I would again stress that when you have genomic testing, it is critical that a physician provide you with a comprehensive overview of the lifestyle changes that can impact your health. Receiving a set of raw genomic data is of no help without genetic counseling by an expert in the area of **Precision Lifestyle Medicine**.

LACTOSE

❉ MCM6 rs4988235

Lactose is the sugar found in cow's milk, sometimes referred to as milk sugar. Lactase is the enzyme produced in your body released from the upper portion of the intestinal tract to break down lactose after partly digested food leaves your stomach. Some people have low levels of lactase resulting in digestive distress (typically abdominal discomfort, gas and possibly diarrhea) when they consume dairy products. This condition is called lactose intolerance and we have long known there is a genetic association, as we see high rates of lactose intolerance in people of Asian descent. Typically, milk, ice cream, and certain cheeses are the main offenders, but portion size also can impact the severity of symptoms. This can be further confusing because someone can instead be allergic or intolerant to milk proteins, whey and/or casein, and get digestive symptoms as a result of their reaction. This type of allergy can be diagnosed with a blood test.

Most people who call themselves lactose intolerant have not had a diagnostic test to confirm it; in fact, most doctors don't routinely run lactose intolerance testing. People just tend to avoid dairy altogether, or at least the specific types of dairy that tend to make them feel sick. Some people have figured out on their own that they need to take commercially available lactase pills every time they eat ice cream or have a meal containing cheese, and they drink Lactaid milk to avoid unpleasant digestive symptoms. If their problem with dairy was an allergy to the proteins, then the lactase pills wouldn't work. I have also found that the overall health of the gut and microbiome can influence how a person responds to dairy consumption.

To definitively diagnose a person with lactose intolerance, physicians order a lengthy lab test that requires the person consume lactose, obviously an unpleasant test. However, again with the utilization of genomics in clinical practice, we can determine whether a patient has the genetic predisposition to have lactose intolerance from a simple saliva sample that also provides a host of other valuable information. The collection of these types of profiles is so beneficial to me as a physician; I don't have to send my patients to a lab for lengthy tests or get multiple vials of blood drawn and have patients risk injury and bruising from the blood collection. We often encounter adults who have a fear of needles or have problems getting their blood drawn, as well as kids who have a great deal of difficulty with having blood testing. There is also a hydrogen breath test that is even more accurate than the blood test but also requires consuming lactose and sitting for several hours, breathing into a machine several times. Although genomic testing is not a substitute for appropriate diagnostic laboratory testing, it can help the doctor determine what tests to run as well as the frequency of testing.

ALCOHOL

✳ ALDH2 rs671

Most of us are familiar with that slight warm feeling that spreads through your body when you have that first glass of wine, or any alcoholic drink for that matter. But for some people, it is much more than a cozy warm feeling, it is a full-on alcohol flush where the face, neck and shoulders turn bright red and they can feel uncomfortably hot, slightly dizzy, get headaches, feel nauseous and even panicky. This has been commonly referred to as Asian Flush because it is genetic and only occurs in people of Asian descent. It is actually quite common in Asians (up to 50%), with the highest frequency in Southeast China. It is present in populations in Japan, Korea, Mongolia, and Indochina. The symptoms can be set off by only a few sips of a drink. People with this SNP often get severe hangovers, as well. This genetic variant codes for an enzyme called mitochondrial aldehyde dehydrogenase, which is necessary in the second stage of alcohol breakdown. When this enzyme is defective, it causes acetaldehyde to build up and results in these uncomfortable symptoms. It also puts the person at higher risk of acute alcohol toxicity.

There is a much greater concern than a red face and neck; this SNP is associated with an increased risk of esophageal cancer likely because of the increased exposure to elevated acetaldehyde levels, which is toxic to the tissues. This risk is elevated when combined with smoking. In addition, this SNP has been linked to colorectal cancer, late onset Alzheimer's disease, osteoporosis, lung cancer and alcoholic liver disease. The only good news is that people with this SNP have been shown to be less likely to suffer from alcohol dependency, which is probably a behavioral mechanism to avoid alcohol because of the somewhat immediate, undesirable physical effects. Having one copy (AG) of this SNP results in slightly less severe symptoms than having two copies (AA), or being homozygous. The risk for esophageal cancer is also less if you have only one copy (AG).

Some of us may get blotchy skin or a mild flush and even hives from alcohol consumption without having this SNP. This is most often due to reactions to other components such as sulfites, preservatives, histamine, grains and other chemicals found in alcoholic beverages. So, if you do get an alcohol flush you cannot automatically assume you have Asian ancestry. But, if this SNP shows up on your genomic profile it is wise to avoid alcohol consumption and make sure your healthcare provider is aware of the health implications of this SNP. It is an important piece of your health history and should be part of your plan for preventative medicine.

CHAPTER 7

#4 VITAMINS:
The Magic Wands of Health

When I first started University (that's what college is called in Canada) my intention was to study biomedical engineering. As a high school student, I was accepted into a summer internship program called Women in Science and Engineering (WISE) and spent my term in an engineering lab at Memorial University of Newfoundland. We were a tight-knit group of girls, all passionate about science, and not only did we get full immersion in a core field of science, we also got to attend lectures given by women who had devoted their careers to science. That exposure to the burgeoning field of biomedical engineering was a tremendous experience and I thought this field was what I was going to do with my life.

Fast-forward 18 months to the middle of my second semester, enrolled at that very same university, realizing that although I excelled in physics, my heart was in the land of the living, biology, specifically biochemistry. About that same time, I also became fascinated with how food affects the body. Fortunately, there was an actual program available that blended the two majors, biochemistry and nutrition. I quickly decided this would be my major and spent the next three years delving into the intricate details of how the foods you consume provide all the tiny elements needed to run the chemical machinery that operates your body. When you study the biochemical pathways that provide for every body function, you become awed by the value of every piece of food you put in your mouth.

Most of you, I suspect, have looked at a food label and perhaps noticed the listing of calories, protein, fat and certainly, carbohydrates, sugar and dietary fiber. But have you noticed other details on that label for vitamin A, vitamin C, calcium, magnesium, iron, vitamin D and riboflavin? These are what we call the "micronutrients," which consist of vitamins, minerals and cofactors. They are called micronutrients because we only need then in very small amounts, in comparison to the macronutrients (fat, protein and carbohydrate) that we need in larger quantities to provide the fuel and form the building blocks for our bodies.

The World Health Organization (WHO), the leading authority for global health policies and education, defines micronutrients as substances that "…are the magic wands that enable the body to produce enzymes, hormones and other substances essential for proper growth and development. As tiny as the amounts are, however, the consequences of their absence are severe." I believe there is no better way to explain the power of micronutrients. I have been studying nutrition for more than 20 years, and working with patients for almost 15 of those years, and I can tell you, without a doubt, that nutrition can completely transform the state of your health. I am not only referring to eating a nutritious diet, but also addressing disease states and dysfunction by identifying vitamin and mineral deficiencies and treating accordingly, as well as prescribing therapeutic dosage levels of vitamins and minerals. I see it every day in my practice.

The amount of scientific evidence that exists in the field of clinical nutrition is outstanding and so extensive that it is hard even for professionals to keep up. Here are just a few medical conditions that are best treated with nutritional solutions:

1. Iron deficiency anemia
2. Iodine deficiency goiter (enlarged thyroid gland and resultant hypothyroidism)
3. Rickets (a skeletal disorder in young children caused by vitamin D, calcium or phosphate deficiency, no longer common in the United States because of the process of adding vitamins and minerals to foods, especially cereals, called fortification)

These are more obvious medical conditions, but every day I find issues in patients that can be addressed with nutritional support, such as using magnesium supplementation for patients with "treatment resistant depression" (meaning antidepressants haven't been very effective for them); vitamin B_6 plus magnesium to relieve symptoms of premenstrual syndrome; and, increasingly, prescribing methylated folic acid (MTHF) for a host of other women's conditions.

Almost all of the vitamins are considered essential, meaning they must come from your dietary intake because you cannot make them in your body. In the U.S. we use a system called the Recommended Daily Allowance (RDA) to list the various amounts of vitamins and minerals needed daily by a generally healthy person to provide *adequate* levels of nutrition. This system categorizes each RDA by age groups and specifies differences needed for genders, pregnancy, and lactation. Its purpose is to provide general guidelines to a broad population but does not account for any preexisting conditions, metabolic problems, and genetic variants a person may have. This is where a healthcare provider trained in clinical nutrition is a key member of your team.

In the category of nonessential vitamins (meaning your body can provide these vitamins in means other than dietary consumption) there are really only a few exceptions. Although these vitamins are synthesized within the body, they are all still required for critical biochemical reactions. Deficiency states in all of these vitamins can still occur:

1. **Vitamin D:** This vitamin is made in your body in response to exposure to sunlight. Unfortunately, we still see high levels of vitamin D deficiency, especially at more northern latitudes, where the level of sun exposure needed to synthesize adequate levels of vitamin D is limited to summer months. Also wearing sunblock and sunscreen, not having enough skin exposed and the level of pigment in your skin all affect the amount of vitamin D you

can make. I'm not suggesting you skip sunblock because the damaging effects of prolonged sun exposure outweigh the risk of low vitamin D. Most people need to supplement this vitamin to stay in an acceptable range. Supplementation ranges widely vary, and, not surprisingly, depend on your genes.

2. **Vitamin K:** Although a form of this vitamin, called vitamin K_2, which is actually a group of compounds called the menaquinones, can be synthesized by the bacteria in your gut, most people still need vitamin K from their diet to achieve the levels necessary for optimal function. This is the reason why there is a RDA set for this vitamin. Vitamin K_2 can also be obtained in the diet from animal sources and fermented foods. On the other hand, vitamin K_1 (also referred to as phylloquinone) is the main type of vitamin K found in the diet. It is plentiful in leafy green vegetables, like kale, Swiss chard, spinach, parsley and broccoli, but also found in tiny amounts in certain oils like olive and canola. If you have ever known anyone on the blood thinner Coumadin or warfarin, you may remember that the person must limit consumption of leafy green vegetables because it can interfere with the effectiveness of the medication. This type of blood thinner acts as what is called a vitamin K antagonist; getting too much vitamin K in the diet will compromise its effect on blood clotting.

3. **Biotin:** Also manufactured by the gut bacteria, biotin is part of the B complex group of vitamins. Most people can make adequate amounts and there is no set daily recommended amount. Like other vitamins and minerals, some people can have higher metabolic needs for biotin, including smokers, or those who have preexisting conditions that require levels beyond what their gut bacteria can provide. Biotin is often added to good multi-nutrient formulas, and can also be found in foods like cooked eggs, avocado, salmon, pork, cauliflower and raspberries.

You would think that in the U.S., particularly in a higher socioeconomic demographic area where I practice, I wouldn't see many vitamin deficiencies, but in fact, I routinely order blood tests for vitamin and mineral levels and commonly find deficiencies in my patients. There are quite a few factors that can explain this trend. A typical standard American diet is generally very low in micronutrients for a number of reasons. Factory farming, the process of producing massive amounts of food, results in soils that are nutrient-depleted as compared to the past. Also, the transportation process and the amount of time it takes to get from the farm to your table have significantly increased, resulting in lower vitamin levels. Furthermore, the higher proportion of processed foods consumed and the move away from a whole foods-based diet result in lower levels of these vital nutrients in the diet.

However, even if you are eating an organic, primarily local, whole foods-based diet, you can still be impacted by genetic variants that affect how well you can absorb nutrients and convert them to the active forms needed for the biochemical reactions in your body. Now, let's examine some of the better known and more serious SNPs that could be keeping you from feeling better than you do right now.

How Your Genes Impact Your Micronutrient Status

As a physician who recognizes the importance of nutritional status as a basic foundation for health, I not only assess the general diet of my patients, I also want to know whether those nutrients are actually getting absorbed, converted and utilized as they should. Blood testing continues to be a necessary part of the evaluation process, but incorporating genomic assessments has taken the process up a notch. Not only can we determine which nutrients we should look for over time, but we can also tell if someone is having problems "activating" or absorbing specific micronutrients, which could be the underlying cause of the complaint the patient describes, such as feeling depressed. If you go online and read about depression, you will find information on a list of vitamins and minerals that could be helpful treating this illness. Often I see that people just start taking all of those supplements. My approach is very different. While I find that foundational nutritional support is helpful, testing allows me to discover which specific nutrients are likely resulting in the biochemical disruption. Armed with this knowledge, I can formulate a targeted approach and the right therapeutic dosage level to address whatever problem was uncovered in the testing process. This is a great place to insert my tagline here: Test, don't guess!

Now let me take you through the SNPs that I routinely examine.

FOLIC ACID

✳ MTHFR rs1801133-1298

This is one of the most widely studied vitamin SNPs; there are actually entire courses you can take on the health impact of this relatively common variant. Having either one or two copies of this SNP incrementally affects the activity of the enzyme that converts folic acid which is found in food and most vitamin supplements into the activated form of folic acid, methylfolate. This is necessary for your body for a range of basic and critical biochemical processes. In my medical practice, I had been using blood testing for many years to assess patients whom I suspected could be having problems with this process. This problem is much more prevalent than I ever expected when I started testing many years ago.

I was first exposed to the "MTHFR mutation" (as it was initially labeled) through a patient case I was managing. A young woman consulted me for depression. She was already under the care of a psychiatrist who had been treating her with standard pharmaceutical antidepressants which didn't seem to be helping that much, so he added another prescription called Deplin. He had ordered a genetic test on her and found she had the "MTHFR mutation." Deplin is a high dose "activated" folic acid or methylfolate intended to increase the effectiveness of certain medications. My patient reported significant improvement after she added the Deplin but was still not feeling as happy as she had in the past, which is why she decided to consult with me. I did identify and treat effectively other underlying causes for her depressive symptoms and she finally got back to leading the happy family life she desired with her husband and three young children.

This is a genetic variant that is quite important; if you are homozygous (have two copies) you have an increased risk of having a stroke and obstetrical complications (if you are female). Because of the fundamental impact of the enzyme affected, I have seen many patients where this SNP has played a major role in a range of health problems from mood disorders and behavioral issues, frequent miscarriages and infertility, polycystic ovarian syndrome (PCOS) to cardiovascular disease. Fortunately, I have also seen miraculous recoveries including women delivering healthy, full-term babies and long-standing depression seemingly cured in a matter of days from taking a supplemental vitamin. And I want to emphasize this is all possible because of genetic testing!

VITAMIN A

* BCMO1 rs7501331
* BCMO1 rs12934922

You may have heard at some point, perhaps from your mother, that carrots are good for your eyes and indeed, this is true. Carrots are a good source of beta-carotene, which is a pigmented phytochemical that is the precursor for vitamin A. Beta-carotene is only found in plant foods and once consumed, it either gets utilized quickly as an antioxidant to protect against free radical damage or is rapidly converted to vitamin A, also called retinol or preformed vitamin A. Vitamin A is important for maintaining healthy tissues like skin, bone, teeth, mucus membranes and, of course, good vision. Retinol is essential for the pigments in the retina and necessary for visual function in low level lighting. Deficiency in vitamin A can result in night blindness and more severe permanent eye damage. Vitamin A deficiency also increases the risk of death from severe infections.

Preformed vitamin A is considered one of the fat soluble vitamins only found in animal-derived foods like whole milk dairy products, eggs, meats, poultry and fatty fish. Non-fortified low-fat and fat-free animal foods have reduced levels of vitamin A; therefore, if you are consuming a very low-fat diet, then preformed vitamin A consumption can also be low. Likewise, if you are consuming a vegan diet, you don't get preformed vitamin A from your food and depend solely on conversion of beta-carotene to vitamin A. This becomes a problem if you have a genetic variant in the BCMO1 gene that codes for the enzyme that converts beta-carotene to vitamin A.

Similar to the MTHFR gene, there are two relatively common SNPs in the BCMO1 gene and the level of activity of the enzyme is correlated to the pattern of SNPs a person displays. For example, someone who has at least one variant in each of the two SNPs can have a 69% reduction in activity of that enzyme. This is a significant conversion problem. The solution here is similar to that of the MTHFR SNP; to reduce the risk of vitamin A deficiency, consumption of preformed vitamin A is essential. When I see patients who have been following a very low-fat or vegan diet, I consider it a priority to assess their nutritional status. Checking vitamin A levels is an easy lab test, but more precisely checking these SNPs to determine if routine monitoring is warranted could save time and money over the long term.

A word of caution when it comes to preformed vitamin A: Because it is a fat soluble vitamin, we store the vitamin A we don't use and this can put us at risk of vitamin A toxicity when you are supplementing with high doses. This is definitely one of the vitamins where too much of a good thing is not good! Furthermore, women who are pregnant or could become pregnant should be very careful with vitamin A. Although vitamin A is necessary for reproduction and development, higher levels of consumption can cause birth defects. Depending on your diet and genetics, your needs will vary. In my practice we use a comprehensive vitamin packet designed specifically for the needs of pregnancy, which contains only a small amount of preformed vitamin A, on the order of 1500 IU per day. (IU means international units and is a measurement standard.)

VITAMIN C

✳ SLC23A1 rs33972313

✳ SLC23A1 rs6596473

Most of you probably associate vitamin C with immune support. I commonly see patients on their first office consultation who have been taking only a couple of nutritional supplements on their own and often one of them is vitamin C to prevent a cold. However, most people are unaware of the more significant role vitamin C plays in tissue growth and repair throughout the body. It is one of the most important vitamins I prescribe for wound healing and postsurgical recovery.

Vitamin C is considered an essential dietary element, meaning we cannot make it on our own and must get it from what we eat. This is why sailors in the 15th to 18th centuries took citrus on their ships to prevent scurvy, a disease caused by vitamin C deficiency that leads to severe periodontal disease. Surprisingly, scurvy still occurs in the U.S., where people have diets lacking in fresh foods that provide vitamin C. A person's genetics also plays a role in how well vitamin C is absorbed. While a person is consuming small amounts of vitamin C from dietary sources, he may lack adequate absorption and be at risk for suboptimal levels of vitamin C and deficiency states.

There are two primary transport proteins that provide for the absorption of vitamin C into your body. One of those proteins is encoded by the gene SLC23A1. This is where this SNP comes into the picture. Variation in this gene reduces bulk transport of vitamin C and has been shown to be associated with a host of conditions involving vitamin C metabolism. People who have SNPs in this vitamin C transport gene can have increased susceptibility to Crohn's disease, aggressive periodontitis, esophageal, stomach and colorectal cancers, preterm births, infections and chronic liver diseases. When I see this SNP on a patient's genomic profile, I monitor for vitamin C deficiency with laboratory testing and advise supplementation. I also now consider this SNP when I evaluate patients who come in with cancers of the GI tract, periodontitis and other conditions that could be related to vitamin C deficiency.

VITAMIN D

❋ GC rs2282679

We have long known the impact of vitamin D on human health and fortunately, the use of vitamin D blood testing has become more widespread. However, we still occasionally get some reports from our patients that confirms health-care providers are not all on board with increased testing. Recently, one of our patient's primary care doctors told him that there was no need to test his vitamin D levels because having that data wouldn't matter anyway. Thankfully, these comments have become less frequent.

The use of blood testing allows us to identify vitamin D deficiencies and suboptimal levels, treat with vitamin D, if necessary, and then retest to determine whether the patient's condition has improved. This process can go smoothly, but without genetic testing, I can really only guess the correct dosage of vitamin D for each patient since individual needs vary so widely. During my pregnancy, I was tested for vitamin D during my routine blood tests. It turns out that I needed quite higher doses than expected of vitamin D. Since having a genomic wellness profile, I have learned that I have a variant that explains what happened during my pregnancy.

VITAMIN E

❋ Intergenic rs12272004

Actually a group of eight compounds, called tocopherols and tocotrienols, vitamin E is a class of fat-soluble vitamins and important antioxidants. Antioxidants are necessary for protection against free radicals that cause damage to your tissues. Considered the most important vitamin E compound, α-tocopherol is also the most studied and is related to the most concerning connection between vitamin E and your genetics. Very severe vitamin E deficiency over many years can lead to problems with walking, muscle weakness, numbness and tingling and even eye damage. Most people (more than 90%) do not get enough vitamin E in their diet, referred to as marginal levels of vitamin E, but some people experience significant problems with malabsorption that can further limit absorption. The effects of these submarginal levels of vitamin E have been the subject of many research studies focused on cardiovascular disease and cognitive health. In one clinical trial, more than 2,000 Finnish subjects were selected to study how genetics affect vitamin levels, specifically α-tocopherol levels. People with a variant in the rs12272004 SNP had lower levels of the most active form of vitamin E. The scientists determined this was likely due to an association with the APOA4 gene that affects circulating lipid levels in the blood that are needed for vitamin E absorption. In layman's terms, if you have this SNP, you may be consuming the recommended daily allowance for vitamin E but your absorption is reduced, thereby lowering the amount of vitamin E you are obtaining from your diet.

Vitamin E is found in mostly plant sources with the highest amounts in nuts and seeds (especially almonds and sunflower seeds), but also in adequate levels in some fruits and vegetables like avocado, spinach, asparagus, blackberries and apricots. When supplementing with vitamin E, it is important to use a supplement that contains α-tocopherol and to consume it with a meal containing fat because it is highly dependent on fat for absorption into the body. Good multivitamin formulas will contain between 30–100 IU of α-tocopherol, which is sufficient to achieve the RDA for adults. You want to avoid higher doses of vitamin E supplementation not only because it is a fat-soluble vitamin, but also because it can have serious consequences if you have certain health conditions. It can interfere with some blood-thinning medications and increase risk of bleeding with nonsteroid anti-inflammatory drug (NSAID) use.

VITAMIN B₆

✳ NBPF3 rs4654748

When I think of vitamin B₆, I always think of Alan Gaby, M.D., one of my medical school professors, a widely known expert in the field of clinical nutritional science. He emphasized the importance of vitamin B₆ in clinical nutrition. Our class even had a joke for tests; if we were in doubt about the answer to one of the questions, we would guess vitamin B₆ or magnesium and it would probably be correct! And for good reason—vitamin B₆ is involved in numerous biochemical processes in the body, including the manufacture of antibodies for the immune system; building hemoglobin, which is the protein that carries iron around in the blood; production of neurotransmitters, amino acid and fatty acid metabolism; and blood sugar, sex hormone and homocysteine metabolism. These are all critical to normal body functions and with low levels of vitamin B₆, symptoms of various problems can easily start showing up.

Although severe vitamin B₆ deficiency is relatively uncommon, low levels of vitamin B₆ can occur fairly often. A

2003 study reported that almost 1/4 of the subjects who did not take supplements had low vitamin B_6 status, while 11% of vitamin supplement users displayed lower levels. Vitamin B_6 is an essential water-soluble vitamin found in both plant and animal sources, although the form of B_6 found in many plant foods, called pyridoxine glucoside, is less bioavailable, which is why we can see lower levels of vitamin B_6 in vegetarians. Alcohol use and certain medications, including birth control pills and NSAIDs, can also affect vitamin B_6 status.

The NBPF3 gene codes for a protein that is part of the neuroblastoma breakpoint family ("NBPF"). This family of proteins is still somewhat poorly understood and within this family of proteins there may be some involvement in the turning on and off of cancer genes. What we do know is that if you have one copy of the rs4654748 (heterozygous CT) you have 1.45 ng/mL lower blood level of B_6, while if you are homozygous CC, the level is twice as low, 2.90 ng/mL lower. When this SNP has shown up on some of my patients' genomic profiles, it can help explain why they are suffering from premenstrual syndrome (PMS), depression, morning sickness or carpal tunnel syndrome, and most importantly, guide me to offer targeted solutions to address their particular health concerns.

VITAMIN B₁₂

✳ FUT2 rs602662

Another essential vitamin that we must obtain through our diets because only bacteria have the ability to synthesize it is vitamin B_{12}. It is considered a water-soluble vitamin; however, it can be stored in the liver for many years but does not result in toxicity (which is why there is no upper level set). The main sources of vitamin B_{12} are meat, poultry, fish and shellfish. Only small amounts are available in eggs and dairy. Therefore, like vitamin A, vegetarians of all types (including those who consume dairy and eggs) are at risk of developing vitamin B_{12} deficiency.

It is estimated that as many as 80% of adults who are vegetarians are at risk of vitamin B_{12} deficiency. I routinely find vitamin B_{12} deficiency when I screen my vegetarian patients with blood testing. But, they are not the only ones who can be affected. As we age, especially after we reach age 50 or so, we don't absorb vitamin B_{12} as well as we did when we were younger. That, combined with a poor diet due to problems with digestion, taste and appetite as well as infections of the stomach and gastritis can limit the amount of B_{12} that can be absorbed. People who have had weight loss surgery (gastric bypass) are affected, as well as those with celiac disease and inflammatory bowel disease. Also, certain medications like those used to treat gastritis, peptic ulcers, and chronic heartburn/reflux including proton-pump inhibitors (PPIs) and H2-receptor antagonists, certain cholesterol medications, some antibiotics and Metformin (a diabetes medication) can all reduce vitamin B_{12} absorption.

It will come as no surprise that genetic variation can also affect your ability to absorb enough vitamin B_{12} to avoid low or deficient levels in your body. The FUT2 gene affects how well B_{12} is absorbed and like many of these SNPs, if you have two copies (homozygous, GG for this SNP) the incidence of B_{12} deficiency is much higher than if you are heterozygous. People who consume vegetarian diets and have this SNP are at very high risk of deficiency and this can help explain why some vegetarians tend to do well with only small amounts of vitamin B_{12} supplementation, while others need to consume higher amounts to keep their blood level of B_{12} in a normal range.

Why should we be concerned about getting enough vitamin B_{12}? It is critical for the formation of red blood cells and deficiency can result in a specific type of anemia called megaloblastic anemia. Low levels of B_{12} can also cause symptoms in the nervous system, including numbness and tingling, generalized weakness, loss of balance, depression and possibly cognitive decline. There is also some evidence that vitamin B_{12} is important in bone remodeling and deficiency could contribute to the development of osteoporosis.

CHAPTER 8

#5 MICROBIOME: Bacterial Genomics—The Microbiota

Unless you're a microbiologist, chances are you never think about the intimate relationship you have with your body's bacteria. The healthy version of this relationship benefits both species, a concept referred to in biology as symbiosis. We basically do nothing on a conscious level other than provide a warm environment with a constant supply of fuel for them to thrive, in exchange for a huge score on our end, when they are in balance. But, when those little guys get out of balance (referred to as "dysbiosis"), boy does that relationship go sour for us!!

All the previous chapters have focused on human genomics but this book wouldn't be complete if I didn't discuss the study of bacterial genomics. While many of you may be familiar with the concept of taking a small capsule containing probiotics, most people do not know that our gastrointestinal tract is host to literally pounds of bacteria! This massive microscopic world has an equally massive impact on human health and wellness. With the advances made in genetic testing, the study of the microbiome has taken huge leaps forward, as well. And that is great news for me as a clinician because now I can order laboratory tests that assess the state of my patient's microbiome.

The microbiome is a term used to refer to the microbial organisms that live in and on our bodies. There are more than 1,000 species that take up residence in our gut, the vast majority of them bacteria. There are actually more of these bacteria thriving inside us than we have cells in our body. Research is now focused on figuring out the specific roles these bacteria play and which species predominantly influence human well-being.

In addition to the genomic testing I order to assess my patient's *human* DNA, I frequently order genomic testing for their *bacterial* DNA, too. Unfortunately, this isn't a saliva test... and you may have already assumed that it requires a stool sample. I send my patients home with a prepackaged stool collection kit and although it sounds unpleasant, it really isn't that bad. After the sample is collected, you package it up in a preaddressed FedEx envelope and pop it into a drop box, thereby avoiding any awkward interaction at the shipping counter when you have to say what the package contains. Patients who dislike blood tests generally don't mind a stool test.

I have been utilizing stool cultures since my days as an intern to test not only for intestinal parasites and infections, like *Giardia*, *Clostridium difficile* and even food poisoning from *Salmonella* and *E. coli,* but also to assess the levels of the two main classes of good bacteria, *Lactobacillus* and *Bifidobacteria*. These types of stool tests are performed by a lab technician, who takes a small swab of fecal material and brushes it across a petri dish containing a specific growth medium, incubates it in a warm environment and then waits for bacteria to grow. The tech identifies the type of bacteria and quantifies the level of growth as NG (no growth), or 1+ through 4+. Although this still provides information about the health of a patient's microbiota, it is only a tiny snapshot of what is really happening in the body.

PCR (polymerase chain reaction) technology allows laboratories to offer a more comprehensive picture of the many bacterial classes within a person's microbiota. The test that I order assesses the most clinically relevant species. When we get lab reports back by using this method, we can determine the level of bacterial diversity that exists in that person's gut, which refers to how many different types of bacteria are thriving in the intestinal environment and their relative abundance. Most of the time when I review my patient's lab reports, the level of diversity is very low. And this comes as no surprise to me because I already suspect that the patient may have an imbalance in his or her gut bacteria. This report can diagnose these types of imbalances and quantify in specific units the levels of bacterial species within the different families that are assessed.

The ability to study these bacteria is relatively new; therefore, as you can imagine, there is much research yet to come in this area. From the studies that have been conducted so far, we have seen that low levels of biodiversity have been associated with various disease states. Now, scientists are looking at the different classes of bacteria to determine which ones and the type of balance may influence specific disease states. For example, there have been a number of studies examining the ratio of *Bacteroidetes spp.* to *Firmicutes spp.*, which has been associated with obesity states. And even more exciting, there is animal research demonstrating that when this ratio is normalized to that of the lean state, it can affect body weight (more on this relationship later in this chapter).

As a healthcare provider, I am faced with the question of what to do with this information. At this point, the technology that is available is still ahead of the clinical evidence and trial data that will eventually provide clinicians with highly personalized treatment protocols to deliver to their patients. However, I am fairly certain this won't come as pharmaceutical intervention. We know that the bacteria in the gut are highly sensitive to environmental factors, like food and medications, of course, but even stress and activity levels can affect the growth of intestinal bacteria. (I have devoted an entire chapter to how your gut bacteria are affected by your lifestyle in part 2 of this book.)

In the meantime, in my experience, there are still effective ways to address the findings from a patient's microbiome lab report. For example, I had a patient who said she had intense bloating and abdominal pain whenever she ate foods containing starchy carbohydrates like grains and potatoes. She had already consulted a gastroenterologist, who could offer her no treatment for these symptoms and didn't think these symptoms were caused by any medical condition. We decided to run the microbiome stool assessment and determined that she had low bacterial diversity from PCR testing, low levels of her "good" bacteria on her culture and even a minor intestinal parasite. These findings required a comprehensive treatment plan. She took an antimicrobial to treat the parasite, started prebiotic dietary support and a multi-strain probiotic that was "strain identified" (probiotic supplements must be from a reliable source and state the genus, species and strain, otherwise you have no idea what you are really taking). Soon after starting this protocol, she noticed a significant improvement in her symptoms. Within a month, she no longer had abdominal pain when she ate the foods that had been triggering her bloating and pain.

In this patient's case, it was obvious that I needed to order stool microbiome testing. She was experiencing symptoms that appeared to be stemming from her digestive tract. However, much of the research being conducted on the state of the microbiome as it relates to human health is focused on dysfunction and disease states outside of the gastrointestinal system, such as diabetes, heart disease, autoimmune diseases, Alzheimer's and even back pain! It's also worth noting the changing concept of where microbes live in our bodies. It was once thought that these tiny guys resided only in our GI tracts, mucous membranes (like the respiratory and vaginal tracts) and skin but now we know that they can even live in our brains! That's why this microscopic world you're hosting is so crucial. Let's take a closer look at this relationship.

Could Your Gut Bacteria be Making You Sick...and Fat?

Mounting evidence suggests that there is a specific profile of gut bacteria that is common to people with obesity, inflammation and insulin resistance. Yes, I guess you could say that your "gut" may be giving you that gut. Your gut bacteria are thought to work like an extra organ, influencing your metabolism. They are even involved in basic functions like the absorption of nutrients. I observed this process up-close in a patient I was treating who had a mild type of arrhythmia which caused her bothersome heart palpitations. Her symptoms were effectively managed with oral magnesium supplementation. However, at the same time, I had also prescribed a probiotic for a mild digestive problem. At a subsequent office visit, she complained that the magnesium she was taking seemed not to be working anymore and she had also run out of her probiotic. We put her back on the probiotic and did not change anything else and her heart palpitations disappeared!

We also have seen multiple studies showing that these bacteria can **cause** and maintain the state of obesity through complex mechanisms like controlling eating behavior (also referred to in studies as "host ingestive behavior") but also affect how you burn calories and the process of fat storage! The older concept of utilizing only the laws of thermodynamics, or what is often referred to as the theory of "calories in vs. calories out," is becoming outmoded. How your body interacts with its environment is much more complex than can be explained by calorie deficit. If you have ever gone on a low-calorie diet with a group, you will know what I am talking about. Everyone doesn't get the same results, even if they're all following the same plan. For as long as I have been in clinical practice, this has been an extremely frustrating situation. Of course, we have known differences in metabolic function that can explain these outcomes. But as we've just spent the last seven chapters talking about genomics, you can see how the old rules are just that and it's time we moved ahead with all the evidence that has emerged.

There are several ways your bacteria can impact your weight. First, they can affect *your* eating behavior. Researchers have cited two potential strategies that bacteria may use to positively influence their own growth:

1. They generate cravings for foods that they specialize in or foods that suppress their competitors.
2. They induce dysphoria (defined as a general state of unease or dissatisfaction with life!) until we eat foods that enhance their growth and metabolism.

We already know that genomic traits can affect our eating behaviors, but this is dramatic evidence that bacterial imbalance in our gut can also influence eating habits.

Now, you've probably heard of your metabolic rate, the rate at which you burn calories, which we already know is influenced by your human genetics. There is now evidence that the bacteria in your gut could slow down your metabolism and result in obesity. Researchers were looking at the effect of an antipsychotic psychiatric medication, risperidone, which is well known to cause rapid weight gain. They examined specifically the effect of the medication on the gut microbiome in an animal model. They found that the drug-induced weight gain correlated with an altered gut microbiome, and even more importantly, they found that this change suppressed the animal's ability to burn calories while at rest. Those gut microbes were transferred to an animal who had not taken the drug and the same effect was observed, causing the researchers to conclude that it is not the drug itself, but the effect risperidone has on the gut bacteria, which results in the rapid weight gain. Researchers have also examined this relationship in children who have been taking risperidone and have observed the same alteration in gut bacteria correlated to a higher BMI. That is why there are far-reaching implications in patients whose guts are imbalanced!

Furthermore, gut bacteria affect how we process fats and especially important, store fats. You might assume that dietary fiber is needed to promote regular bowel movements by providing bulk for the stool. You may have heard that fiber can reduce the risk of colon cancer and help prevent heart disease. But now we know that dietary fiber is extremely important to the health of our gut bacteria. The microbes in our gut ferment the fiber that we consume and produce short chain fatty acids (SCFAs) from it. These SCFAs have been studied extensively and have been shown to protect against obesity. Exactly how does that work? Well, you may guess that it has to do with the balance between fat burning and storage. Butyric acid and propionic acid are two SCFAs that have been shown to inhibit the body's ability to make fat and stimulate the breakdown of fat. This is exactly the pattern that promotes improved lean body mass.

If you spend any time looking at fitness magazines or websites, you will be familiar with headlines of "fat loss" instead of "weight loss," or "the top 12 ways to burn fat." In fact, the evidence in this area is really starting to show that all of these techniques and lifestyle changes are linked to the effect on the gut bacteria. There are countless studies and more emerging frequently that all point to the role of gut bacteria in fat metabolism. An interesting recent study even questioned the relationship of altered gut microbiota in pregnant woman possibly resulting in excessive weight gain from fat storage. More research needs to be done in this area to determine whether addressing gut bacteria could play a role in management of weight gain during pregnancy.

At this point, the evidence is very clear that the microbiota plays an integral role in obesity and maintenance of a lean body mass. We know that there are specific bacterial profiles found in the gut of obese subjects when compared to lean subjects, specifically the population of *Akkermansia spp.* and ratio of *Bacteroidetes* and *Firmicutes.* These families appear to be of particular importance as they make up the bulk of the gut bacteria. So can you change the bacterial profile to that of a lean subject and induce fat loss? The short answer would seem to be yes! Although there have not yet been any published studies on humans, two very recent studies conducted in animal models treated with specific plant extracts and probiotics resulted in a shift in the bacterial profile to that of the lean animal and associated significant reductions in loss of fat tissue. In lay terms, we could say that if you get your gut bacteria back into balance, it could start showing on your waist line!

Sounds too good to be true? Well, addressing dysbiosis is much more complex than just taking an antimicrobial plant and some over-the-counter probiotics. Once the microbiota gets back into balance with a precision-level treatment plan, there must be a concerted effort at maintaining healthy gut flora. We know that the bacteria in the intestinal tract are influenced by many factors.

The composition of the diet is one of the primary influences on the microbiota. A study published in 2015 in the *World Journal of Diabetes* examined the effect of a diet rich in carbohydrates, whole grains and vegetables, with no animal fat or protein or added sugar for the treatment of type 2 diabetes. The results showed significant improvements in fasting blood glucose, hemoglobin A1C, lipid profile, body mass index, body weight and blood pressure, which they suggest involves alteration in the gut microbiota. More and more studies are showing that the effect of healthy dietary habits is not just from the way the nutrients affect metabolism, but from the complex interaction with the gut bacteria.

More to the Story than Just Obesity

Although, most of what I have just explained was focused on obesity and metabolism, as I noted earlier in this chapter, the state of health of our entire microbiome has broad-reaching effects on every body system. There are already books devoted entirely to this subject area. My intention is to show you the importance of your microbiome and how the lifestyle choices you make every day can impact the health of the microscopic world in your body. Let me share a few more clinical cases where addressing microbial balance has been an effective treatment approach in instances where you probably would never have guessed it could make a difference.

A few years ago, a 30-year-old female patient came to me for chronic sinus infections and allergies. For at least the last five years, she had been treated about three to four times per year with antibiotics and sometimes an oral steroid pack for sinus infections. She was also receiving weekly immuno-therapy injections for her environmental allergies. Overall, she told me she hadn't been getting better, constantly felt congested and had postnasal drip. She had also developed some digestive problems. Initially, I started a treatment protocol to help stabilize her immune system using nutrients including quercetin, vitamin C, vitamin D and fish oil. She also started eating a diet that reduced systemic inflamma-tion. On the probiotic side, because she had taken so many courses of antibiotics, I advised her to start a higher dose balanced probiotic containing equal amounts strain-iden-tified *Lactobacillus acidophilus* NCFM and *Bifidobacterium lactis*. Over the next few months, she noticed a significant improvement in her level of daily congestion and did not need treatment for a sinus infection. She also said her diges-tive complaints resolved. This is a general immune-balancing approach that I have seen provide tremendous benefit to my patients. However, this patient continued to have a low level of mucus production, especially in the winter (a common complaint I see here in Connecticut) that never seemed to resolve completely despite many treatments tried over the years...that is until a new probiotic strain became available. This probiotic contains two strains of bacteria that have been shown clinically to support the health of the sinuses and nasal passages, specifically. Within a week she noticed a dramatic improvement and continued to use the probiotic into the spring. She started again in the late fall when her "winter nose" started acting up again and noticed the same results. This treatment has worked with many of my other patients who I have advised to take these same probiotics for similar reasons.

Another common complaint I often hear from women is about recurrent urinary tract infections and vaginal infections. These women are often treated with rounds and rounds of antibiotics which can wreak havoc on the gut microbiome (more on that in part 2). Previously, the integrative approach was to provide treatment with the probiotic containing *Lactobacillus acidophilus* NCFM, and often adding general immune-supporting formulas containing nutrients like zinc, vitamin C and vitamin D, and sometimes immuno-modulating herbs as well. Now, we can better treat these conditions with a "precision-biotic," another blend of strain-identified probiotics that have been clinically shown to support the urinary and vaginal tracts, thereby effectively reducing the recurrence of these types of infections. This has been such a blessing to many of my female patients.

These are very specific examples of the value of using precision probiotic therapy. However, there are many other treatment approaches to influence the health of the microbiota, most importantly dietary therapy and the use of specific compounds found in nature. The research in this area is very active and there are even ongoing efforts to "engineer" probiotics with pharmaceuticals. The implications for therapy for treating chronic conditions and disease states through addressing gut health are vast.

As I stated earlier, in my practice, we have found micro-biome assessments to be an invaluable tool to gain a better understanding of digestive health. We have observed that many of our patients who come in for weight management may also have digestive problems (and then this testing comes in handy), but you can now see how knowledge of what's happening in the microbiome, regardless of whether you are having tummy troubles, can be useful. This is definitely another area where **Precision Lifestyle Medicine** using SNP analysis and microbiome profiling enhances our treatment approach.

Part 2

EPIGENETICS

How Your DNA is Affected by Your Environment

CHAPTER 9

An Introduction to the Science of Nurturing

S ince researchers cracked the code of the human genome about a decade ago, we know of the role genetics plays in whether we will develop certain diseases and disorders. A growing body of evidence has confirmed the connection between our genetic profile and the likelihood we will have adverse health outcomes, including cancer, heart diseases and depression. As a result, some people choose to get genetic testing to find out whether they have genes that are linked to a higher risk for diseases or disorders. However, other people question the value of learning their genetic risk profile if they cannot change their genetic predisposition.

Before I answer, let me remind you that in the first part of this book, I frequently explained how the environment—from what we eat to how much we move our bodies, our exposure to pollutants and even how we deal with stress—impacts the expression of our genes. For example, research shows that younger women, more frequently express breast cancer genes. This means that something about our environment is causing those genes to be expressed into the manifestation of a disease state.

This is also true for celiac disease. You may carry the genes for celiac disease but may never express the actual disease. I have seen patients who have a parent with celiac disease and the patients have the genes that could get turned on, but these people don't have celiac disease. What determines whether the gene variant is turned on? Most of chronic disease is the result of a complex interplay of our genes and our environment. In the case of celiac disease, proposed mechanisms include everything from physical stressors like a recent surgery to infections in the gastrointestinal tract.

Our Genes are Not Our Destiny

Not that long ago, everyone assumed that our genes were our destiny, and that we could not influence them. Even today, many people fear having genetic testing because they feel they would rather not know their eventual "prognosis." When I talk to my patients about having wellness genomic profiles (not testing for cancer genes), they sometimes are concerned that if they find out they have a problem, they can't do anything about it, so they'd rather not know. By now, if you've read this book carefully, you understand that is not the case. The good news is that we are not at the mercy of unchangeable genes. Although our underlying genetic makeup may carry vulnerabilities and a higher risk for disease, the outcome is not carved in stone, and neither are our genes.

Research has shown that at a given time, only a small part of our genome is active. Furthermore, your genes can be "turned off" and "turned on." As I just mentioned, this gene expression can be dynamically altered by your environmental conditions. The study of this phenomenon is called epigenetics and suggests that we may have more control over our genes in terms of disease prevention than had been previously thought.

Nature v. Nurture...or is it Nurture Shapes Nature?

Twin studies are widely used to distinguish between the influence of genes and the influence of the environment on human traits. Some of these studies, such as the landmark "Minnesota Twin Family Study" conducted from 1979 to 1999, examined identical twins who were separated at birth. Although some results have demonstrated the power of genetics, there were twins who, despite the fact that they had the same genetic profile, showed different behavior, personality and health outcomes. Previously this was a classic nurture (environment influence) versus nature (genetic inheritance) debate about which external factors after conception were thought to influence health directly. However, the science of epigenetics provides more insights into this debate, suggesting that in identical twins the same genes may express differently due to different external influences. It appears nurture can shape nature through epigenetics.

Do Not Skip this Part!!

Understanding the mechanisms by which all of this occurs involves some highly technical language, and similar to genomics, there is also a "language of epigenetics." This requires you to understand some rather complex concepts so bear with me while I try to simplify the basics. This subject is critical to your health so try to get comfortable with the terminology.

First of all, a naturally occurring chemical process called **methylation of DNA** acts like a tag that shows whether a gene is turned on or off. The tags are chemical compounds called **methyl groups** that are on or attached to the DNA. Now this part can get a little confusing because when DNA gets methylated, or "tagged," it can turn genes off by blocking the expression of a gene, but it could also mediate some gene expression, thereby turning the gene on. **DNA methyl-transferases** and DNA deaminases are enzymes involved in turning genes on and off. This process is influenced by external factors like your diet, activity level, exposure to chemicals and even stress. As I have already said, while you can have a gene for a particular condition, it might not be turned on because of the lifestyle choices you are making.

These methylation patterns on a cell's DNA are referred to as an **epigenome**. Methylation patterns differ by tissue type, meaning a fat cell's epigenome will appear different from a muscle cell's epigenome. Furthermore, the epigenome of a normal cell doesn't look the same as a cancer cell.

DNA methylation is a pretty stable process, except in cancer, which is where most of epigenetic research to date has focused. Typically, DNA in a cancer cell has a lower level of overall methylation, also called **global methylation**, compared to a normal cell, except at sites that control the cancer cell's growth. There are also genes called **tumor suppressor genes** that limit the growth of cancer cells, and these genes can get "silenced" through the process of DNA methylation. Researchers have been trying to develop tests for much earlier detection of cancers, such as colon cancer, based on these characteristics.

Okay, that's the toughest part. I hope I haven't lost you. Go back and read the previous two paragraphs a few times to make sure you understand the basics before moving on. There are also other processes or **epigenome modifications** that influence how your genes are expressed, including **acetylation of DNA** (which typically turns genes on), methylation and modification of **histones** (these can change rapidly during a cell cycle) and **chromatin modeling**. The interplay of all of these very complex reactions consequently affects the expression of genes associated with normal and pathological processes. Through these mechanisms, disease risk is either increased or decreased, which in turn affects longevity, the aging process, as well as your overall health and well-being.

Passing On Your Genes and...Your Epigenome?

Your genes were passed on to you from your parents and you cannot change them. But, you can change and choose the external factors that determine whether your genes are expressed or not, and thus, whether your gene sequences activate the development of a disease to which you may be genetically predisposed. Scientists are finally learning how nurturing affects our DNA.

Would you believe me if I said that you might be able to pass on the changes in your epigenome to generations to come? It turns out this appears to be the case. Although expression patterns of your genes occur throughout your lifetime, research shows that your epigenome is, in fact, heritable, just like your genome. There is scientific evidence that the stress from physical traumas, psychological insults (including bullying) and socioeconomic stress (including famine) affects the epigenome and can be passed through multiple generations. There is even a study showing that olfactory exposures can be passed along, referred to affectionately as "heritable memories" prompting one headline to read, "Have researchers cracked the case for past life memories?"

CHAPTER 10

Behavior, Stress, and Hormones

Since the 1970s scientists have known that DNA requires epigenetic instructions on how to be read and expressed, and more recently we have learned that those instructions can come specifically from DNA methylation and histone modification (which we just reviewed in chapter 9). When researchers were first working out the details of how gene expression was affected, no one understood at which points in life epigenetic influences played a role. Initially, scientists believed only influences in utero played a role in our genetic expression. Today, however, we know that our genes interact with our environment throughout our entire lives, and even sustain the impact of stressors from previous generations. In this chapter, I will focus primarily on how various forms of stress (including psychological and physiological) can impact your DNA and therefore the expression of your genes.

117

Your Time in the Womb

It turns out that those nine months you spent in your mother's belly could also have been the first time you were exposed to stress. It is well-known that toxic exposures like alcohol consumption result in a defined condition called Fetal Alcohol Syndrome, and other toxins and certain prescription medications can also cause birth defects from their effects on fetal growth and development. Interestingly, maternal stress in pregnancy could also result in symptoms that may not be immediately evident and may not appear until adulthood.

One of the most important studies on this topic was the Dutch Famine Birth Cohort Study. This study of 2,414 people evaluated the role of extremely low-caloric intake of women who were pregnant between November 1943 and February 1947 and the influence on their babies' lifelong health. During the most intense period of the Dutch famine, or *Hungerwinter*, between November 1944 and May 1945, women had intakes of as low as 400-800 calories per day, less than a quarter of the recommended intake for adults.

The most important findings of this birth cohort study were that people who were in utero during the *Hungerwinter* were more likely to have diabetes, cardiovascular disease, elevated cholesterol, obesity and a variety of other health problems later in life. At the time it was thought that poor maternal nutrition led to increased disease susceptibility later in life, and that the timing of exposure of undernutrition was important as well. Different adverse health outcomes occurred depending whether the exposure was in early, mid, or late gestation. These results suggested to scientists that adverse exposures in utero were possibly the most critical exposures.

Now, with the advent of advanced technology used to study DNA, we have a much better understanding about how these observations are evidence of epigenetic changes in DNA. In a landmark animal study published in 1997, scientists at McGill University found that the offspring of more affectionate and attentive mothers had reduced response to acute stress than did the offspring of mothers who exhibited less grooming behaviors toward their offspring. When the results were first released, the researchers did not know the reason for the reduced stress response was due to epigenetic influences like methylation; they believed the result was due to maternal care.

A later study, published in 2004, concluded that the outcomes of reduced stress response were, in fact, due to epigenetic influence through histone modification and DNA methylation. The authors concluded, "An epigenomic state of a gene can be established through behavioral programming, and it is potentially reversible." While in the case of the rat model, greater methylation of the hippocampal cells in the offspring brain was associated with greater stress response, the same is not true for methylation of all cell types. Rather, seeking "optimal" methylation is the goal–not necessarily more or less.

This pattern ring is also thought to apply for other traumas. Offspring of Holocaust survivors were thought to be more predisposed to certain conditions in adulthood because of the changes in child-rearing that could result from being the parents subjected to extreme treatment and conditions.

But now, scientists believe the explanation comes from a sort of "biological memory" because their parents' DNA was marked from the trauma. Because researchers have identified transmission of epigenetic changes across multiple generations, there may be an association between the biological memory that may exist in African-Americans whose ancestors were slaves and their increased rates of heart disease, diabetes and obesity. No research to date has been done in this area, but it warrants investigation.

The Formative Childhood Years

The "in womb" experience is not the only period that has a profound impact on your health status. Insults to our systems affect gene expression from childhood throughout our adult lives. Childhood stress also plays a role in genetic expression and the risk of developing chronic disease later in life. The ACE Study, currently underway with the Centers for Disease Control and Prevention and Kaiser Permanente, is attempting to develop an index to quantify risk of adult illness based on ACE, or *adverse childhood experiences*, including abuse, violence and other traumas a child may have experienced early in life. Since the late 1990s, more than 70 studies have shown a strong link between a high "ACE score" and a variety of chronic diseases, including ischemic heart disease, depression, and obesity, as well as greater health risk behaviors like alcohol and drug abuse. While our experiences in utero play a role in determining our lifelong health, the ACE Study suggests there are still important windows of opportunity, such as childhood, in which stress and trauma can influence disease pathology and health outcomes. Other studies support these findings and show that epigenetic changes in childhood have a far greater influence on later health outcomes than do exposures later in life.

Animal studies have provided evidence supporting changes in the DNA methylation as a result of stressors. The team at McGill University in Montreal, Canada, looked specifically at epigenetic changes from childhood experiences in humans. They studied a group of men and found DNA methylation patterns that matched up with either early childhood family wealth or poverty. Socioeconomic stress in early life can have a long-lasting effect on the genome. This same group of researchers also studied two groups of Russian children—those raised with their biological parents and those raised in orphanages. Not surprisingly, they observed increased methylation patterns in the group raised in orphanages, and of particular note were changes in methylation of genes related to brain growth and neural function.

And the Stress Goes On...

The impact of stress on epigenetic expression in our bodies depends on a variety of factors such as the type of stress, age, gender, hormonal state, ability to control stressors and previous experiences. That's a very important point which I discuss with my patients every day; while we have little control over many of the stressors in our life, we do have the power to control how we *respond* to stress. It's also essential to understand that the stress response is critical to body function. Acute stress helps us to stay alive and alert, while long-term or chronic stress can alter neurons in the brain, and the brain's structure and function, which can reduce cognition and alter behavior, even leading to neurodegenerative diseases. It's up to you and crucial to your well-being that you focus on addressing how you're responding to stress in your life. The most effective method, both in terms of feeling better and cost, is meditation. In the Precision Health Program in part 3 of this book, I explore meditation in detail.

We also know that changes in protein production in the brain due to long-term stress can be passed from one cell to the next, as well as being passed on to offspring. The most significant finding in epigenetics is that "stress-induced epigenetic changes can persist long after the stressor has ended and can cause functional changes in the brain." Epigenetic changes caused by stress are highly interdependent and are caused by a variety of factors, such as the exact DNA sites, timing and non-epigenetic factors.

The Hormonal Cascade

Often, prolonged periods of extreme stress can lead to what we term "burnout," which may involve symptoms of depression such as exhaustion or lack of interest in activities that you once enjoyed. One consequence of prolonged stress is adrenal fatigue, which is caused, in part, by prolonged elevated levels of the hormone cortisol. Beyond the psychological and physiological symptoms of burnout, researchers at Johns Hopkins University School of Medicine found that long-term exposure to cortisol in mice resulted in greater expression of anxious characteristics, with alteration of methyl groups on a gene with known links to mood disorders in humans. Similar findings have been observed in humans, with individuals suffering from anxiety found to have higher levels of global DNA methylation. A 2010 study also showed that elevated levels of cortisol have an epigenetic influence on the development of cardiovascular disease in older adults. In that study, the subjects with the highest levels of cortisol had a five-fold increased risk of dying from heart disease.

Are those enough reasons to just chill out?! This stress stuff is a really big deal. I like the social media post I have seen circulating recently saying, "If you don't have 20 minutes to spend outside enjoying nature, then you should actually spend two hours outside enjoying nature!"

CHAPTER 11

DNA and Your Lifestyle: Diet and Exercise

You know you should follow a healthy diet and exercise. You've heard that for years...so what? Frequently, my patients haven't been motivated enough to adopt healthier habits. However, when I explain the direct link between the risk of developing diabetes (like a mother), the person usually has a light-bulb moment and is more motivated. You have the power to turn off those genes, simply by what foods you choose to eat and whether you exercise. The choices we make multiple times a day are essentially providing a stream of signals to our genes. The next time you are faced with the choice between an apple and a muffin, you might be more likely to choose the apple. The old concept of "I'll eat this now and make up for it later with a healthy dinner" is outdated. When you start thinking about all the opportunities you have to promote a constant stream of health, instead of the old "counting calories" method, your entire approach to wellness can change. Next, let's look more closely at how your exercise level and your diet, and even your grandfather's diet, affect your epigenome.

"I am What I Eat" in another Perspective

Personal choices determine the environment we live in and our exposure to environmental factors. Among others, these choices include the food we eat and how much we exercise. Diet and nutrients may influence epigenetic mechanisms critical for gene expression. These mechanisms are dynamic and the effects of environmental factors such as nutrition are reversible. Nutrients that are considered important in changing or modifying epigenetic processes are methyl-donating nutrients, such as folate, choline, betaine, B vitamins and methione, and their supplementation is especially critical during fetal development or before conception. This fetal programming has been demonstrated in agouti mice. All mammals have a gene called Agouti. When the Agouti gene is on or not methylated, the fur color of the mice is yellow and the mice are obese and prone to diabetes and cancer. Yellow-furred Agouti mice normally give birth to yellow offspring. However, there is evidence that the Agouti gene is turned off if the gene is methylated. If a pregnant agouti mother is exposed to a diet with methyl-donating nutrients, the Agouti genes in the embryo will be methylated and the offspring will be brown and healthy, having a lower disease risk. The positive effects of maternal diet on children's health have been demonstrated in human studies as well. Likewise, nutritional influences (including alcohol) on the epigenetic processes and health outcomes, such as diabetes mellitus, obesity, cardiovascular disease, cancer and longevity, have been described in adulthood. The current knowledge on nutrition, epigenetic mechanisms and health, however, is still limited, and mechanisms responsible for epigenetic health effects are complex.

Could My Grandfather's Diet Actually Affect My Health?

Not only does the environment affect our genes, it may also have an effect on gene expression in our children and grandchildren, according to a Swedish study by Bygren and Pembrey. Epigenetic influences can be transmitted from one generation to another, according to the scientists who investigated well-kept historical records of annual harvests in a small Swedish community. This study showed that drastic changes in the availability of food during childhood had an effect on next-generation health. But the influences were gender-specific. Grandsons of boys who had experienced an abundance of food when they entered puberty died earlier than those of boys exposed to famine. The opposite was true for granddaughters of women who had experienced famine in the womb or just after birth. They had an increased risk of early death. So starvation in women during fetal or early post-natal life seems to lead to worse health outcomes, in contrast to men who had an abundance during pre-puberty. Apparently, during development, epigenetic mechanisms capture environmental information about nutrition and this is passed on to later generations. These transgenerational epigenetic mechanisms were also demonstrated in studies that investigated the famine during the last years of the Second World War in the Netherlands. Children of women who were pregnant during the famine in 1944–1945 (especially when they were in the early stage of their pregnancy) had a higher risk of adverse health outcomes in later life.

How Superfoods Affect Your DNA

The quest for the latest and greatest superfood always makes a great story. We are literally bombarded by headlines about these foods, and they often come with bizarre names along with a promise to transform your health. What exactly is superfood and what does it have to do with how your genes are expressed?

A superfood is not a medical term; it is a label coined to describe a specific food that contains compounds, often in higher amounts, that are beneficial for your health. These compounds could be basic nutrients like vitamins, minerals or fatty acids such as omega-3s found in salmon. More often, though, these popularized superfoods contain health-promoting plant compounds that we refer to as phytochemicals. You may have heard that red wine is good for you because it contains resveratrol. This is a great example of a phytochemical in a food...well, wine, but same difference, right? Resveratrol has been the subject of countless research studies; of particular interest is its potential role in longevity, although it seems that would require extremely high doses, like drinking four bottles of wine, which obviously is not safe. Nevertheless, resveratrol has other health benefits such as its antioxidant activity, which protects against DNA damage. Resveratrol is in a class of compounds called polyphenols. It appears this class of phytochemicals is critical for health promotion. They are found in many herbs, fruits and vegetables, but also in tea, coffee and chocolate. Chocolate, wow!

I believe chocolate is one the best superfoods available in nature. It sits at the top of my list along with green tea, blueberries, coffee, broccoli, soy and turmeric (the main spice in curry powder). The health benefits of dark chocolate, especially from a specific class of compounds referred to as cocoa polyphenols in science circles, have received a great deal of media attention for two reasons. This concept is supported by scientific theory and we all know that people love chocolate and like how they feel when they eat it. Because this is an extract of the cocoa bean, like any other plant, chocolate contains many compounds. I'll simply focus on how chocolate affects our epigenome.

Researchers in Spain examined the effect of consuming chocolate extract in a group of adults who had cardiovascular risk factors. They divided the subjects into two groups for two weeks, feeding one group chocolate while the other ate a diet controlled for the same amount of calories without the chocolate. Then, they tested the DNA of the people studied by taking a sample of their white blood cells to determine the level of global methylation—remember, in most cases more methylated (hyper) is bad, while less methylated (hypo) is good. They found that people who were eating chocolate had lower levels of methylation. They even looked at the effects of chocolate on people who had SNPs in their MTHFR and MTRR genes that are involved in supplying methyl groups; the positive effects were not as pronounced in the subjects who had the SNPs. This means that chocolate actually affects the expression of our DNA in a beneficial way. For many years I have been telling my patients to eat a couple of ounces of dark chocolate, which contains more than 70% cocoa powder, because of the beneficial cardio-vascular benefits. Now, I have even more reason to prescribe chocolate! How's that for a prescription?

Not only has chocolate been shown to affect the epigenome, studies have demonstrated that bioactive components in some of the other foods we frequently eat can also affect methylation patterns of our DNA. Another polyphenol compound, epigallocatechin-3-gallate (EGCG), found in green tea, can undo hyper-methylated DNA, which also increases the expression of genes that inhibit, or suppress, tumors, thereby playing a "good guy" role in cancer. For all you coffee drinkers, don't worry; I'm not going to suggest you switch over to green tea. It turns out that some of the compounds found in coffee, like caffeic acid and chlorogenic acid, also inhibit DNA methylation. I, personally, drink both coffee and green tea.

Sometimes patients come to me saying they have heard soy is bad for them, and they should avoid eating it. This is not founded in science; I can't even begin to tell you how many beneficial effects can be attributed to the compounds found in soy foods in terms of cardio-metabolic health. In the epigenetic arena, a specific compound called genistein (which falls in the isoflavone category) has been shown to have a similar effect as the EGCG in green tea, increasing tumor suppressor gene activity and reducing the global level of methylation of DNA. I encourage my patients to only eat non-GMO or organically produced soy whenever possible.

The list of foods that impacts the epigenome is quite extensive. Curcumin is found in turmeric, which is the predominant spice in curry powder. This bright orange pigmented compound has been labeled "a potent hypomethylation agent." In plain English, that means it is has the same positive effect as the compounds found in green tea, coffee and soy. Research has also shown that bioactives in cruciferous vegetables like broccoli and leafy greens, black raspberries, apples and grape seeds all have beneficial effects on the epigenome. And pharmaceutical companies are developing hypomethylation agents. But why not let food be your medicine? With the multitude of plant compounds, there is not yet one single agent that can match the health benefits of consuming a diet rich in phytochemicals.

Moving Your Body Can Shake Up Your Genes

As you can imagine, if diet has such a profound impact on our DNA, then the level of exercise or how active we are must also play a role in how we express our genes. In fact, exercise can directly influence how we express our genes. Through physical activity, a high risk gene expression for disease can be changed into a low risk or healthy gene expression. A sedentary lifestyle, on the other hand, has been linked to unfavorable epigenetic changes and to adverse cardiometabolic health outcomes. Strikingly, the influence of exercising on gene expression may be immediate, influencing genes after a single workout.

How the epigenome appears varies by cell type, meaning the epigenome of a fat cell can differ from that of a muscle cell. Up until very recently only a few human tissues had their epigenome examined, but that trend is changing. One of the first studies to look at the epigenome of fat cells in humans involved an experiment examining the effect of exercise on the DNA methylation pattern in these cells. In the study, 23 Swedish men who had a baseline low level of activity had samples of fat tissue taken from their thighs before they started a six-month exercise program. The program consisted of a one-hour spinning class per week plus two other one-hour sessions of aerobic activity led by a certified instructor. At the end of the program, another fat tissue sample was collected. DNA methylation patterns were then measured in the fat cells. The results showed significant alterations in global methylation patterns after the men had been exercising for six months, and they were further able to associate these changes in their epigenome with fat metabolism. Previous research has shown similar changes in the epigenome pattern in skeletal muscle cells induced by exercise.

Knowing Our Epigenome: A Promising Challenge

As I've repeated throughout this book, it turns out that genes are not our fate. Evidence is starting to show that we can control our destiny and even that of our children and grandchildren by influencing the expression of our genes with healthy nutrition and adequate physical activity. By adhering to a healthy regimen and by appropriate lifestyle adjustments we can encourage the expression of genes that lead to an overall improved state of health.

In the beginning of this book, I talked about the value of genetic testing to find out your genetic risk profile. However, just knowing your genes might not be enough to take the right decisions to prevent disease and maintain health. In addition to genetic testing, epigenetic mapping and being aware how your genes are being expressed by diet and exercise may be essential to changing your lifestyle. Both genetic testing and epigenetic profiling may help to prevent disease and help you personalize your lifestyle according to your specific medical needs. Finding out whether you have a higher epigenetic risk profile may help you make important health decisions to lower your risk and get better health outcomes. Also, in some cases, genetic testing and profiling can lead to early disease detection, and consequently to early treatment.

Because of its dynamic and reversible character, this new epigenetic concept offers major possibilities for the future of disease prevention and disease treatment. Our genes are willing and able to unlearn bad influences and implement healthy ones. While the genome changes slowly, through the process of random mutation and natural selection, the epigenome (DNA with attached compounds) can change rapidly and switch—as in response to signals from the environment. However, the challenge will be finding the evidence and identifying nutritional factors, appropriate diet and type and amount of exercise that can positively influence the epigenetic process.

CHAPTER 12

DNA, Aging, and "Inflammaging"

(Get ready for a little more technical terminology.)

It's not just you looking in the mirror at the lines around your eyes. People are getting older and living longer. Adults over the age of 65 make up about 8% of the global population, and that trend is projected to increase in the coming years. By 2050, this demographic group will double to 16% of the population. That's a key reason that the anti-aging segment of the beauty product industry is highly profitable.

It is important to distinguish life expectancy from *healthy* life expectancy because many elderly adults spend their final years suffering from disease and disability. The concept of having more "life in your years" is referred to as *compression of morbidity*. Most people would prefer not to spend their later years suffering from chronic illness and in a nursing home. The majority of conditions that impact older adults are considered age-related chronic diseases, such as heart and lung diseases, diabetes, Alzheimer's and cancer. These health issues are likely to coexist, elevating the risk of frailty, death, disability and reduced quality of life in the elderly.

Are We Aging or "Inflammaging?"

It's quite simple—we age chronologically. This process of aging is associated with changes in the adaptive and innate immune system with essentially a loss of immune function. As a result, aging is characterized by a chronic, low-grade state of inflammation, which is the result of an imbalance between inflammatory and anti-inflammatory responses to all our external stressors. I often describe the very act of getting older as a mild inflammatory state. The concern here is not just the appearance of more lines on your face; chronic inflammation is a risk factor for morbidity and mortality in the elderly. It's also a major determinant of global aging and longevity in general. *Inflammaging*, as it is called, and its accompanying loss of immune function helps explain the increased incidence in age-related diseases, since most chronic diseases share an inflammatory origin.

Although important in longevity and health, inflammation is not the only factor impacting aging. Genetic predisposition certainly plays a role. There are certainly people age 90 or older who didn't have particularly healthy lifestyles. But as you know, your genes are not the complete story; nurture or environmental factors are also important.

Diet can have a major effect on the aging process, specifically through a process called glycation. This occurs when circulating blood sugar levels are too high and sugar molecules end up getting attached to protein or fat, thereby impairing bodily functions. In my practice, we routinely run a lab test called hemoglobin A1C which indicates how much glycation is present. These levels will be very high in diabetics who don't control their blood sugar very well. This is why so many metabolic systems are affected in diabetics, and results in an advanced rate of aging. When my patients ask how they can slow down the aging process, I first recommend a lower glycemic diet. So, it seems that we are aging and inflammaging at the same time and these processes are very much interconnected.

Our Age, Written in Our DNA

Some people live to be 100, while some die long before they reach that age. Women live about four years longer than men, and some elderly suffer a variety of diseases while others don't. There are so many individual differences that researchers have been trying to evaluate which factors contribute to longevity and health. As I mentioned, we do know that our aging process is genetically determined. Over the last few years, scientists have identified several areas in our genes that have been linked to longevity and age-related diseases. Inflammaging also has a genetic basis. It determines the body's threshold of pro-inflammation and the sensitivity to inflammation, which are both important for the development of diseases. Apparently, we are genetically programmed to age, and it appears even to inflammage. Furthermore, DNA damage and change, which occur over time, also contributes to how we age. One of these DNA changes occurs in telomere length. Let me explain because telomeres are thought by some to be the holy grail of aging.

Telomeres are sequences of the DNA at the end of the chromosomes that protect the genetic information in the chromosome. When cells divide (which is part of normal development) the telomeres shorten, allowing cells to divide without losing genes. When the telomeres are too short, cell division is no longer possible without losing the necessary information. Because the cells are no longer functioning, they die. The enzyme *telomerase* protects and repairs the telomere from shortening. But, although it gives some protection, telomeres still shorten over our lifespan. The *telomere shortening* process is thus associated with aging, and more specifically, longer telomeres have been linked to healthy aging.

Is Inflammation Genetic?

By now, you are aware that we have genes related to aging and longevity, and we have genes that are associated with various health outcomes. We also have specific genes that control our immune and inflammatory processes. Genetic studies have identified numerous genes and biological pathways important in inflammation and inflammatory diseases. Through this, not only do genes and genetic variants impact disease risk, they are also linked to disease progression and severity, and consequently to aging. In the last few years several genes have been discovered that encode inflammatory molecules, such as cytokines and coagulation factors. Research has shown that species that live longer have more genes that encode molecules that help to control inflammation. Also, genes have been identified that control inflammatory responses by activating telomerase, repairing telomeres and protecting it from shortening. Inflammatory genes are also likely to collaborate and these complex networks affect disease risk and longevity. Research over the past 20 years has brought to light a completely different understanding of the inflammatory response.

Turning Off Inflammation...and Aging

When I talk to my patients about inflammation, I often talk about putting out the fire. The general approach to inflammation has been focused on treatment with anti-inflammatory agents, whether synthetic or naturally derived. It turns out that is not the entire story. We now know there is another phase of the inflammatory process called resolution of inflammation which involves specialized pro-resolving mediators (SPMs) which basically complete the inflammation process. The way I approach inflammation in my practice has changed dramatically, whether I'm treating obvious inflammatory conditions like rheumatoid arthritis as well as "silent inflammation" and inflammaging.

Another way to address inflammation and inflammaging is through the concept of epigenetics. We know that we can affect our genes. Recent evidence shows that beyond the genetic basis, there is a role for epigenetic mechanisms, such as methylation, in inflammaging as well. DNA methylation in different genes, including inflammatory genes, changes with aging, and influences vulnerability to inflammaging in the elderly. Studies in identical twins have shown that global and gene-specific methylation may be affected by environmental influences during life. In these twins, methylation levels were similar in early life, but showed marked differences as the twins aged, particularly if they were separated early in life. But we knew this already. Genes can be altered by external factors, including genes involved in inflammation.

Studies have also shown that DNA methylation of immune genes, including interleukin-10 and interleukin-16, changes with age. Likewise, gradual age-related changes in methylation were found in tumor necrosis factor, a cytokine involved in systemic inflammation. These alterations may be important in the mechanisms of age-related increased systemic inflammation and the development of inflammatory diseases that come with older age. Besides DNA methylation, microRNAs may play a role in the development of age-related inflammation. These microRNAs are small molecules that can turn on and off genes. A number of microRNAs have been reported to play a role in modulating inflammatory responses and in the aging process. The bottom line: scientists are demonstrating that environmental factors such as lifestyle can impact the aging process, and again, your genes appear not to be your destiny.

Epigenetics: The Window of Opportunity

It's unknown to what extent inflammaging or longevity is controlled by epigenetic events in early life. However, thanks to recent research, it has become more evident that epigenetic plasticity in adulthood may provide a window of opportunity for epigenetic-based lifestyle interventions to alter epigenetic changes later in life and positively affect inflammation, aging and lifespan. Epigenetic signatures, including DNA methylation and microRNAs, can be affected by a wide spectrum of environmental factors including smoking, exercise, psychological stress, sleep, drugs and nutrition. Consequently, inflammaging and inflammatory diseases may have strong switch-like epigenetic origins in gene regulation.

Diet has been shown to affect the development of inflammaging and prevention of age-related diseases and aging, itself. Nutritional epigenetic research is now concentrating on anti-inflammaging dietary interventions. Several dietary components have been described to modulate epigenetic expression of target proteins and key genes involved in inflammaging. The proper diet may therefore protect us from many age-related diseases. But diet may just as likely predispose us to inflammatory age-related health issues. A recent study of more than 800 people in Boston showed that higher intake of omega-3 fatty acids correlated with lower methylation levels of interleukin-6 gene, which was associated with lower interleukin-6 expression in the blood, having a positive effect on inflammation. This study provided evidence on the potential of epigenetic therapy for inflammaging and disease prevention. This was supported by other research work that showed that in colon cells the polyphenols quercetin and resveratrol (back to superfoods) influence the expression of microRNAs related to oxidative stress, an important contributing factor in inflammation. Also, quercetin supplementation has been shown to increase hepatic miRs involved in inflammation. Remember, these polyphenols are found in foods such as tea, coffee, olive oil and red wine and their anti-inflammatory effects and use in prevention of age-related diseases such as cancer and cardiovascular diseases have been demonstrated.

So through epigenetic mechanisms, our environment can influence gene regulation, and consequently the development of age-related factors, aging and longevity. But can inflammation itself affect our genes, and inflammatory processes that are related to them? It can. Chronic inflammation and oxidative stress are thought to accelerate telomere shortening, consequently influencing longevity and health.

One Day We Will Fully Understand Aging, Won't We?

Inflammaging is a pathological phenomenon that provides more insights into our understanding of longevity and age-related chronic diseases, functional decline and frailty across the life course. Understanding the molecular and epigenetic basis that controls age-related inflammation is crucial to understanding whether treatments that modulate inflammaging will help in healthy aging. Epigenetic mechanisms have been suggested to modulate processes related to inflammation and nutritional epigenetic interventions seem to offer opportunities to prevent the development of inflammatory age-related diseases, improve health, well-being and longevity. However, we are still only at the beginning stage of understanding the complex relationships among epigenetics, nutrition, health and aging.

CHAPTER 13

Living in Our Toxic World

Imagine a sandy beach, clear blue water. Imagine a forest, dunes, an idyllic village in Asia or a country meadow. Pictures like that trigger feelings of nature, purity and health. You don't think of chemicals when you imagine places like that, do you? But there is no place on earth without chemicals. Chemicals are all around us, and some—like oxygen and hydrogen—are essential for humans and other living creatures. Many other chemicals, however, are harmful to your health. These harmful chemicals or toxins are numerous, and we are exposed to various toxins every day that could have considerable health risks. Toxic chemicals can be found in the air that we breathe or the water we drink. They are even in the products that we eat or use. With air pollution, soil contamination by pesticides and chemicals like bisphenol A in baby bottles, endocrine disruptors in skin care products, mercury in fish, and lead in computer monitors, the world has become a toxic place.

All I Need is the Air that I Breathe

Little did The Hollies know about the possible dangers of the air they breathed back in 1974 when they released their hit song. Recently, air pollution has been recognized as a major environmental risk to health, a global killer. It is the number four risk factor for death worldwide and, in 2012, one in eight total global deaths (about seven million) was caused by air pollution exposure. Air pollution is estimated to reduce life expectancy by almost nine months. Air pollution refers to certain gases, dust, fumes or odors in harmful amounts. It includes, among others, exposure to particulate matter, such as sulphate, nitrates, ammonia, sodium chloride, black carbon and water, and to ozone, nitrogen dioxide and sulphur dioxide.

Both outdoor and indoor air pollution have been linked to health problems, including diminished quality of life. In the Western world, we don't think much about indoor pollution, but in middle- and low-income countries, indoor cooking and fires pose a major health threat to children and women. However, even in the U.S., we should be more concerned about toxic exposures inside our homes that can arise from off-gassing of volatile organic compounds (VOCs) in new carpets, paint and even new upholstered furniture.

Outdoor air pollution in both cities and rural areas was estimated to cause nearly four million premature deaths worldwide per year in 2012. By reducing air pollution levels, countries can lower the burden of disease from stroke, and heart disease, lung cancer, chronic respiratory diseases and acute lower respiratory infections, mainly in children. Recent data shows that air pollution exposure also affects the central nervous system, including autism and schizophrenia. Reducing air pollution could save lives, prevent disability and increase quality of life for many people. When you hear arguments about climate change and pollution, the conversation usually becomes a heated political and financial issue, but ultimately we are talking about human lives and preventable illness.

Heavy Metals

Not only is air pollution a major health hazard, exposure to heavy metals can cause extensive dysfunction in multiple body systems. These metals are widespread environmental contaminants which have been linked to kidney damage, cancer, cardiovascular disease and neurological diseases. Surprisingly, when I run a comprehensive nutritional panel, which contains a blood test for heavy metals, I can often find higher than "normal" levels of some of these metals which could be contributing to my patient's conditions.

The main threats to human health from heavy metals are associated with exposure to lead, cadmium, mercury, arsenic and nickel. Cadmium exposure can come via batteries, smoking and food. We are exposed to mercury through food, especially fish, and dental amalgams. Normally, "low mercury" fish consumption does not necessarily lead to adverse health effects, but high consumption has been associated with neurological damage. Pregnant women are advised to consume lower mercury-containing seafood no more than twice per week because of the risk posed to the fetus. Major sources of lead include air, food (mostly through packaging) and water. Also, lead emissions have contributed considerably to air pollution and consequently to higher lead exposure. Children are at particular risk for adverse health outcomes due to lead, which is why pediatricians routinely run a blood test on young children to examine their lead levels and identify possible exposures like old chipping paint or water supply. Of great concern to me (especially with respect to the recent Flint, Michigan, water crisis) is a recent paper published in the prestigious scientific journal *Nature* which showed how lead exposure during pregnancy negatively affected the epigenome, and even worse, this damage could be passed along for multiple generations. We are primarily exposed to arsenic through drinking water and food, such as rice and poultry, but also from treatments applied to lumber in the past. Nickel is another carcinogenic metal. It is widely used in industry, but also in jewelry and coins, and chronic exposure has been related to an increased risk of disease.

Although the adverse health effects of heavy metals have been studied and acknowledged for some time, exposure to them still exists and is even increasing in some areas, especially in less developed countries. Obviously, reducing the levels of the toxic metals in our environment is key, as well as providing public education on ways to lower the level of exposure. Once I find elevated levels of heavy metals in a patient, I first try to identify the source of the exposure, i.e., too much sushi grade tuna could elevate mercury levels.

Then, I focus on supporting innate detoxification through upregulation of the enzyme metallothionein, which helps clear heavy metals from the body. Also, there are genetic tests that can be run to identify impairments with detoxification. Obviously, this is a serious problem so you should consult a health care provider who is knowledgeable and experienced in managing heavy metals and detoxification.

Smoking: Creating Your Personal Toxic World

Since the 1970s, when the risks of smoking weren't widely discussed, a lot has changed. Backed by scientific evidence, it is now common knowledge that smoking is bad for you. Smoking cigarettes continues to be the number one cause of preventable death. It has been linked to lung cancer, chronic lung diseases (COPD), heart disease and stroke. All of these diseases are among the top 10 causes of death and disability worldwide. In the U.S. alone, about 30% of deaths from heart disease are attributable to (passive) smoking. Children are especially sensitive to the negative health effects of passive smoking (inhaling cigarette smoke from the environment). Even more striking, smoking is linked to about 90% of cancer cases.

Although the risks are clear and people are aware of them, nearly one-fifth of the adult population continues to smoke. Taking into account that cigarette smoke contains more than 5,000 chemical compounds, it's clear that by smoking you create your own toxic world, and one for others (especially the children) around you. Many of the chemicals in cigarette smoke are harmful to your health, such as arsenic, cadmium and tar. Also, the carbon monoxide and nicotine you inhale when smoking have adverse health effects. But diseases and death caused by smoking are preventable and millions of people successfully quit smoking every year.

Your DNA on Toxic Chemicals

We know that these chemical toxins, whether they are in air, food, water or even in beauty care products, are a health risk. But how? What pathways are responsible for these adverse health outcomes? One of the mechanisms is through our DNA. Chemical toxins, including secondhand smoking, pollution, toxic metals and toxic chemical additives to food, have been shown to damage our DNA. These toxins can drive human mutations. Mutations are subtle changes in the pairing of chemical letters of DNA—adenine, cytosine, guanine and thymine—that produce new cells with different traits than their ancestors. An example of cellular mutation due to environmental damage is called our old friend the SNP (single nucleotide polymorphism) or a copying error in the DNA. Japanese researchers have shown that, during the DNA replication, environmental chemicals lead adenine to pair with thymine, rather than with its correct partner, guanine. Also, environmental factors, such as irradiation and ultraviolet (UV) light from tanning beds and too much sun exposure, can affect cell metabolism and damage DNA bases, causing lesions leading to cell death. These mutations can influence disease and death, but perhaps even more alarming, this toxic environment can also temporarily affect our genes, through epigenetic mechanisms.

Epigenetics and Environmental Exposures

A rapidly growing body of evidence shows that chemical toxins affect your health through epigenetic alterations, including changes in DNA methylation, histone modifications and microRNA. Environmental pollutants can cause hyper-or hypo-methylation of specific genes. You may remember that this is associated with changes in gene expression and therefore affects risk of disease. Particularly impacted are the histones, with chemically induced modifications, which are another type of epigenetic abnormality. These modifications affect the way our genetic material is packed. Consequently, it influences DNA reading regions and gene expression. Histone acetylation and methylation are the most common reported histone modifications induced by environmental exposure. Environmental chemicals also affect our genes through oxidative stress and inflammation. (I described the mechanisms of inflammation and epigenetics in the previous chapter.) Furthermore, environmental exposure can affect global DNA methylation. Studies show that alterations in global DNA methylation may lead to genomic instability and an increased number of mutational events. A large portion of methylation sites within the genome are found in repeat sequences and "transposable elements" which are associated with several diseases, including cardiovascular disease, diabetes and cancer.

The most frequently studied environmental pollutants in relation to epigenetic changes are heavy metals, including arsenic and nickel. Evidence shows that exposure to arsenic affects global DNA methylation and methylation in specific genes. In the human body, arsenic is processed into other metabolites. This process requires S-adenosyl methionine (SAM), which is a universal methyl donor that determines DNA methylation. It has been shown that arsenic exposure leads to SAM insufficiency, and may cause global DNA hypomethylation. This hypothesis is further supported by studies in mice and rats which show that arsenic exposure was associated with global hypomethylation in liver DNA, and alterations in gene-specific DNA methylation in tumor suppressor genes and estrogen receptor genes. Likewise, animal studies have shown that exposure to nickel causes DNA hypermethylation and induces gene activation. Also, nickel may cause gene silencing via histone modifications. Other environmental studies show that exposure to particulate matter by air pollution leads to epigenetic alterations. A recent study conducted in Boston reported that particulate matter derived from vehicular traffic is also associated with changes in DNA methylation.

Preventing Toxic Damage to Your DNA

I find the statistics on chemical exposures alarming. Approximately 24% of diseases are caused by environmental exposure and every year there are more than 13 million deaths worldwide due to environmental pollutants! In a screening study conducted in the United States, 150 environmental chemicals were found in blood and urine, providing evidence on the extent to which environmental chemicals may impact our health. These adverse health outcomes, however, can be averted by preventive measures, and biomarkers reflecting exposure to environmental pollutants can help predict the risk of future disease. Also, epigenetic mechanisms and modifications due to environmental pollutants may provide further clues about how diseases occur. Understanding and preventing disease, and counteracting adverse health outcomes through epigenetic mechanisms, are promising options in fighting disease and death induced by environmental chemicals.

So, the Hollies were right after all, with some additions: All I need is the air that I breathe. And the water I drink and the food I eat…as long as they're free of toxic chemicals. And now that there's a greater understanding of the toxic environment and its impact on health, my concern for environmental toxicity is elevated. At this point, no matter how clean you live in the modern world, it is impossible to avoid toxins. The best approach to reducing your exposure level is by eating chemical-free food and using nontoxic cleaning products, beauty supplies and household products. I also encourage consuming detoxification supportive foods like half a lemon squeezed in water and broccoli/cruciferous vegetables daily. Keeping live plants in your home can help to reduce indoor air pollution. In my clinical practice, we routinely utilize metabolic detoxification programs or ongoing detox support to help address chronic health problems. For some people, testing SNPs related to detect one's ability to detoxify is warranted and depending on the findings, specific supplementation can be of additional benefit. Again, I emphasize the need to consult a physician who has expertise in this field.

CHAPTER 14

Your Outside World Meets Your Inside World: The Microbiota

The balance of bacteria in the intestinal tract is affected by many factors. You probably accept that medications like antibiotics can have a significant detrimental effect on your flora, but it may come as a surprise that lifestyle factors like diet, exercise and even the amount of stress you experience can also influence the gut microbiota. From the moment you're born, not only you but the bacteria in your intestinal tract are responding to the environmental cues of your surroundings. Your microbiota appears to be reacting to everything you do from eating chocolate, going outside for a run and even all those stressful deadlines!

Perhaps you have heard of the benefits of taking a probiotic to support your digestive health. The World Health Organization defines probiotics as "live microorganisms which when administered in adequate amounts confer a health benefit on the host." Evidence supports the role of targeted probiotic use to ameliorate symptoms of specific conditions and support an improved state of health. There are also many ways to influence your microbial community through lifestyle choices that you make every day. The goal of this chapter is to explain how environmental exposures alter your microbiota and, more importantly, help you learn how to gain control over your gut health.

Your Entry into the World

What is referred to in the world of medicine as "delivery mode," whether you were born by C-section or vaginal birth, is reflected in the composition of your gut microbiome. Even by six weeks, the microbiota in the intestinal tract of an infant delivered by C-section looks significantly different than the bacteria in the gut of a baby delivered naturally. Longer term studies are now underway to determine exactly how long that pattern persists, and more importantly, what impact that has on the development of diseases. In part 1, I reviewed how we already know that specific patterns of the gut flora have been associated with conditions including insulin resistance and obesity. What we don't know yet is how major an impact delivery mode has on your state of health. Even more important is whether there are interventions that could be taken early in life that could normalize the gut bacteria to that of a baby born vaginally, and then again determine how that affects a person later in life.

Researchers at the Geisel School of Medicine at Dartmouth College in New Hampshire Children's Environmental

Health and Disease Prevention Research Center have been conducting precisely these types of studies. They have a large-scale ongoing program enrolling newborns into a microbiome project collecting data about the babies' health and their gut microbiota. The data that is being collected will provide valuable insight into the specific roles of the various species of microorganisms in the development of disease, and the impact of early exposures in life on long-term health. It's a very exciting time in this field.

Early Feeding Patterns in Infancy

It has been well established that babies who are formula-fed have a significantly altered pattern in their gut microbiome compared to babies who are breast-fed early in life. The message that "breast is best" is founded on a mountain of research demonstrating the very clear health benefits of breast milk, everything from immune function and allergies to obesity and insulin resistance. But the message that breast milk supports a healthy microbiome has not really gotten much press. And, I predict that the long-term health benefits will end up being related to how specific compounds found in breast milk affect the gut bacteria.

We already know that there are indigestible carbohydrates found in breast milk that appear to support the growth of our good bacteria. From an evolutionary standpoint, having carbohydrates that can't provide a fuel source for human metabolism doesn't make sense. But, we now know that these carbohydrates feed gut bacteria, which is important for our health. Therefore, that these carbs are in breast milk now does make sense when we consider the unique benefit our gut bacteria provide us. In fact, we have started using a medical food in my clinical practice that contains this very type of carbohydrate to support gut health in adult patients who have imbalances.

These carbohydrates are not found in commercial baby formulas, which researchers suspect may be part of the reason why the microbiota of babies who are formula-fed look so different. You may wonder what the microbiota of babies who are fed a mix of breast milk and formula is. Well, until very recently there wasn't any data on the pattern of the microbiome of babies who were "in the middle." It is a relatively common practice in the U.S. and Canada to supplement breast-fed babies with formula for medical reasons like poor weight gain or low milk supply. But some families may decide to supplement for convenience, assuming their babies are still reaping the benefits of breast milk. Therefore, a research study (again from the team at the Geisel School of Medicine at Dartmouth) examined that very question with a group of 102 newborns.

During their first six weeks of life, 70 babies were exclusively breast-fed, six were exclusively formula-fed and the remaining 26 received a combination of breast milk and formula. The question was whether the microbiomes of the babies on the combination feeding look more like formula-fed, breast-fed or somewhere in the middle. What do you think? You probably guessed "middle," and that was my guess, too. Surprisingly, it turns out the microbiome of breast-fed babies supplemented with formula look more like the exclusively formula-fed babies! The researchers say they have much more work to do to better understand the long-term impact of their findings and, most importantly, explore ways to address alterable early life risk factors.

Your Gut on Drugs

Taking just one course of antibiotics couldn't do that much harm to our gut bacteria, right? It turns out the type of antibiotic you take as well as the preexisting state of microbial balance could play a role as well. Time and time again, I have seen patients who have digestive complaints that developed during or following a course of antibiotics. I have seen everything from severe cases of antibiotic-induced diarrhea to daily abdominal pain in kids, lingering for months after they have taken just one course of antibiotics for an upper respiratory infection.

Fortunately, targeted probiotic therapy and sometimes adding a microbiome-supportive diet have resulted in complete resolution of symptoms in the majority of these patients. However, it is far more difficult to address the patients who have digestive symptoms that have developed after rounds and rounds of antibiotic therapy, in some cases for many years, taken for conditions like recurrent sinus infections, ear infections, urinary tract infections and as a prophylactic with dental work. The concern about antibiotic resistance by public health officials led to a newly issued set of strict guidelines for physicians in the U.S. regarding appropriate clinical use of antibiotics.

In Europe, similar problems with nonevidence-based use of antibiotics as preventative care in dental care and prophylactic postsurgical use prompted researchers to conduct a randomized placebo-controlled trial in two centers, one in the United Kingdom and one in Sweden. They studied patients who were prescribed antibiotics, first taking samples at baseline and at specified time points through the course of 12 months. A number of very important observations were made. First, they generalized that the use of certain antibiotics can have severe and long-term impact on the microbiome. They observed that that the oral bacteria bounce back more quickly but the intestines get hit hard. They measured a metabolite called butyrate produced by your healthy gut bacteria which is known to lower inflammation and oxidative stress in the intestines. When the subjects took a single course of the antibiotic clindamycin, decreased butyrate-producing bacteria were measured for four months afterward. People who took a single course of ciprofloxin, however, still had decreased butyrate-producing bacteria for 12 months. So a full year afterward, the microbiome were still disturbed! That occurred only after one course of antibiotics. What happens with repeated courses? I routinely see patients taking three to four courses of antibiotics during fall-spring season for recurrent sinusitis.

Not only can antibiotics affect the health of your gut bacteria, many other drugs can upset the intestinal balance. You might remember how I mentioned in part 1 that the psychiatric medication risperidone results in a specific pattern of bacterial dysbiosis that causes weight gain. A host of other medications such as corticosteroids, acid-suppressing medications, birth control pills, hormone replacement therapy and NSAIDs (nonsteroidal anti-inflammatory drugs) can also mess up your gut bacteria. Interestingly, some medications, including those for cancer, actually require intestinal flora to be effective.

What You Feed Your Little Ones Matters

The consensus for many years was that when we consumed food, it was digested, absorbed and then used as fuel for our cells to carry out the host of biochemical reactions that keep our bodies running smoothly. Now, we know that there is a lot more to this pathway, particularly before the food actually gets absorbed in the intestinal tract. The interaction of dietary compounds with our microbiota has become a very hot topic for scientists and of particular interest to physicians like me.

You may remember from part 1 how an imbalance of gut bacteria can affect your metabolism (specifically, higher *Firmicutes* levels were associated with obesity). Well, it appears the balance of your gut bacteria is responsive to what you are eating. This issue is a bit of the chicken and egg question: Is it cause and effect or a reflection of a diet that causes certain levels of bacteria to flourish?

There has been a tremendous amount of research in this area. We know that varying levels of macronutrients (carbs, fat and protein) can impact the gut. The types of carbohydrate and fibers and even very specific types of food can influence the broad spectrum of microorganisms found.

A recent study examined the effect of following two different diets over the course of a year in an obese population. Twenty men who were part of the Coronary Diet Intervention With Olive Oil and Cardiovascular Prevention study

(CORDIOPREV, which is an ongoing prospective, random- ized, opened, controlled trial in patients with coronary heart disease) were put on either a Mediterranean diet (Med diet) containing 35% fat, 22% monounsaturated or a low-fat, high- complex carbohydrate diet (LFHCC diet) with 28% fat, 12% monounsaturated. The results showed that their microbiomes looked significantly different. Men who followed the LFHCC diet had increased the *Prevotella* (thought to be more inflam- matory) and decreased the *Roseburia* genera, whereas those on the Med diet showed the complete opposite pattern. They also found that the abundance of *Parabacteroides distasonis* increased on the Med diet while *Faecalibacterium prausnitzii* increased on the LFHCC diet. These two diets, often recom- mended to fight heart disease, produce completely different patterns in the microbiome.

We have also seen research showing diets with high protein with carbohydrate restriction associated with negative changes in the microbiome. Groups of bacteria negatively impacted include *Bifidobacteria spp.*, *Roseburia spp.* and *E. rectale* from the *Clostridium* group. As well, high protein-moderate carb and high protein-low carb diets also resulted in the lower levels of beneficial butyrate.

The negative impact of a Western diet with respect to levels of *Firmicutes* (remember, this one is important in obesity) can be modified with both fat reduction and carbo- hydrate reduction—but exactly how much is still debated. There are also multiple studies showing negative effects of high-fat diets, but the details of exactly which types of fats are still being examined.

Then, there's the effect of *prebiotics*, which are defined as "non-digestible food ingredients that beneficially affect the host by selectively stimulating the growth and/or activity of one or a limited number of bacterial species already resident in the colon, and thus attempt to improve host health." Prebiotic fibers include oligofructose and inulin, which have also been the subject of quite a few research studies.

Oligofructose in the diet can alter the gut microbiota by increasing the abundance of *Bifidobacterium spp.* and has even been associated with reduced gut permeability (sometimes called "leaky gut") and metabolic disorders. Another study shows that inulin-type fructans (ITFs) could buffer the effects of a high-fat diet.

The takeaway message is that it really does matter what you feed your bacteria. The details are still unclear and newer studies are examining what effect consumption of specific foods like casein, cultured yogurt and grain sorghum has on the microbiome. In fact, let's take a closer look at one of my favorite foods—chocolate!

Chocolate, a Feast for Your Gut

That title may sound bizarre, but I am again referring to the good bacteria in your intestines. These little critters apparently love to eat the larger beneficial compounds called polyphenols that are found in cocoa powder. Polyphenols are also found in tea and dark berries and we have known for quite some time that there are beneficial vascular effects and possible effects on weight loss/management attributable to these compounds, but the exact mechanism was not fully known. It appears we are getting much closer to understanding this full process.

These polyphenols are not broken down in the stomach or by digestive enzymes. They are processed by gut bacteria, and considered to be a prebiotic, which is a substance that can fuel your good bacteria. Consumption of cocoa polyphenols is associated with higher levels of *Lactobacilli* and *Bifidobacteria* (these are the good guys, the ones you want more of) and lower *Clostridia* levels (the ones you definitely don't want to feed in your gut). This beneficial microbial profile has been associated with reduced inflammation (using C-reactive protein as a marker), improved immune function and lower triglyceride levels. Also, patients who have irritable bowel syndrome (IBS) have demonstrated lower *Lactobacilli* and *Bifidobacteria*, and higher *Clostridia* profiles on stool testing.

As a doctor who treats patients with chronic inflammation, digestive disorders and metabolic syndrome, this gets me really excited. I can write science-based prescriptions for chocolate!! Life is beautiful. Now, I am not talking about a milk chocolate bunny here. This research has been conducted using cocoa drinks containing high amounts of polyphenols. I swapped out my morning shake for a plant-based French vanilla one containing 100mg of cocoa polyphenols #morningbliss!!

Exercising Your Gut Bacteria

Knowing that the overall composition of your microbiome is affected by the method in which you were born, feeding patterns early in life, antibiotic use and your current dietary habits, you can probably guess that your activity level could also influence the balance of these microbes in your gut. And it turns out that exercise matters, as well. There are a number of research experiments that demonstrate the beneficial effects of exercise but I would like to share my favorite one with you. This study is near and dear to my heart because my husband was a rugby player. (In fact, that's how our paths first crossed.) So when I see "rugby" and "microbiome" in the same sentence, I'm excited.

To study the effects of exercise on the diversity of gut bacteria, gastroenterologist Fergus Shanahan, from the University College of Cork, chose a group of extreme athletes—rugby players. (Probably not the first choice of researchers here in the United States, but in Ireland, rugby is *the* extreme sport!)

Researchers analyzed stool samples from 40 players during their training season and compared them to a healthy group of men matched for weight and age. Exercise levels were tracked and the subjects also completed detailed food frequency questionnaires. The researchers also assessed various inflammatory and metabolic markers, plus creatine kinase (which is elevated in extreme exercise). The results showed that the rugby players consumed slightly more protein than the other group (22% vs. 15% of their calories) and, of course, had higher creatine kinase, but they also had significantly higher microbial diversity in their gut bacteria—22 different phyla, to be exact. The researchers demonstrated that a greater variety of specific groups of bacteria is associated with lower risk of some chronic health conditions.

The study concluded, "The results provide evidence for a beneficial impact of exercise on gut microbiota diversity but also indicate that the relationship is complex and is related to accompanying dietary extremes." This is more evidence that a healthy diet, exercise and lifestyle result in improved health, but it makes me wonder if the ultimate source of wellness is coming from a thriving, healthy microbiota. The jury is not in on that, yet.

And Finally...Stop Stressing Out Your Bacteria!

Yes, that's right; your stress response can alter the levels of the microbes in your gut. Knowing the broad-reaching effects of stress on the body, I'm really not surprised to see concrete evidence in support of how damaging stress can be on the microbiome. We all know stress is bad and we should be doing a better job controlling our exposure to stress and how we respond to stressors. But do you really know the extent to which stress can affect you and even your children? I won't go on and on citing study after study but I will review one study that had a profound impact on me. I believe it is an excellent example of how important it is to reduce your stress response as much as possible.

Researchers have already demonstrated that chronic stress in pregnant women suppresses vaginal immune response, resulting in increased vaginal infections. They then set out to determine if these changes in vaginal flora could translate to an altered microbiome in offspring and whether that could potentially affect metabolism that could predispose to mood disorders and neurodevelopment conditions, like autism spectrum disorder.

This was an animal study in which pregnant mice were stressed through non-pain-inducing methods including 60 minutes of fox odor exposure or repeated cage changes. First, the scientists identified loss of *Lactobacillus* in the vaginal flora (also known as a dysbiotic microbial community). Next, they measured outcomes in the gut microbiome of the stressed mother's offspring and found higher levels of *Clostridium* and *Bacteroides* in the males, but not in the females. They also measured altered hippuric acid levels in these male offspring. Hippuric acid is a metabolite shown in previous studies to be associated with neuropsychiatric disorders including autism spectrum disorder and depression. Researchers went a step further to look at changes in amino acid profiles in the brain and found there were consistently lower levels of serine, histidine, asparagine, glutamine, glutamate and glycine only in the male offspring of the stressed mothers. As a result of these findings, the researchers concluded that maternal stress may "contribute to reprogramming of the developing brain that may predispose individuals to neurodevelopmental disorders." That's a pretty big deal, and I will just leave it at that.

I am a firm believer in the concept that "when you know better, you can do better." And this is the primary reason I wrote this book. By taking the time to educate my patients, I know they will be better equipped and motivated to do whatever it takes to maintain an improved state of health. Progress, even in small steps, is really what makes you feel better. And this brings me to the last part of this book: the program. This is a blueprint for the steps to better health that I have developed from both my clinical and personal experiences.

PART 3

THE PRECISION HEALTH PROGRAM

Healthy living is not a series of specialized diets that you follow, it is the lifestyle you dedicate yourself to on a regular basis that impacts the expression of your genes.

I hesitate to use the word "program," because this is really a way of living, your lifestyle. It's not a "30-day fix" or a "six weeks to be beach body-ready" plan. Making the best choices that are aligned with your individual genetics and optimizing the expression of the genes associated with health and resilience is the basis of this lifelong program. Since a personalized lifestyle program is based on your genetics, if you choose to have a genomic profile completed, you can then further customize your program. Much of the information presented here will address the epigenetic side of things, as well as establishing vital lifestyle practices that can improve your success when faced with choices about what to eat and what types of exercise to perform.

I have designed the Precision Health Program with three key areas of focus: (1) mind, (2) body, and (3) spirit. Of course, these are all intimately connected but, for the purpose of clarity, the information is presented in distinct sections. I have also included checklists at the beginning of each section so that if you are already practicing elements of this program you don't need to focus so much on that material and can move on to some other areas where you may need more support.

In the Appendix, I include a test called *Health Status Questionnaire (HSQ)* that will help you complete a personal assessment of your level of wellness. This is a questionnaire I use in my practice as a global measure of health status. I encourage you to complete the HSQ before you start implementing the lifestyle changes outlined in the *Precision Health Program*. Then repeat it again after one month, and again at one- to three-month intervals (you can go to my website www.drjenstagg .com to get new HSQ forms). It's important that you don't look at your previous assessment while you are filling out your next HSQ.

A word of caution: This is meant to be a general guide-line for healthy living and is not a substitute for medical care. Certain metabolic problems and health conditions must be taken into account when you're making changes in your lifestyle. I recommend working with a licensed healthcare provider trained in functional medicine or naturopathic medi-cine to help you select the lifestyle program that best fits your individual needs.

CHAPTER 15

MIND

Many philosophers have said that when you get your mind right, everything else will follow. I completely agree with this statement. That's why when a patient comes to me with a primary concern of weight management and is also depressed, I focus first on improving psychological health while recommending simple lifestyle changes. As I have already said, the biggest challenge with adopting a healthy lifestyle is maintaining it or, as doctors say, "compliance." Many people generally know what they should be doing to achieve better health and well-being. Now, with the help of genomics, doctors can advise people on what precisely they should be doing to maintain a state of optimal health. But it is just the "doing" that can be the game-changer.

I don't want you to think that having a healthy mind is only for the purpose of making dietary and exercise choices that promote health. Maintaining a great state of mental health is a key pillar of vitality. We know that depression is an independent risk factor for cardiovascular disease and not simply because depression is causing the person to make poor lifestyle choices. A healthy mind is just as important as a healthy body. Mind and body are inseparable. The health of the body impacts the health of the mind and vice versa. (That's why the health specialty is referred to as mind-body medicine.)

PHP: Mind Checklist

- I am routinely sleeping seven to eight hours every night and waking refreshed.
- I always have a positive outlook on life.
- I am grateful for all that I am and all that I have.
- I write in a gratitude journal every day.
- I spend time outside in nature almost every day.
- I treat myself with love and respect.
- I respond in a calm manner instead of "reacting" in stressful moments.
- I feel balanced in daily life and not overwhelmed by routine daily activities.
- I practice creative visualization.
- I regularly use relaxation breathing.
- I feel comfortable using techniques to stabilize my emotions when I am encountering negative feelings.

SLEEP

This is an area of your life that should not require much effort but you do need to plan and prepare. Sleep is a vital process during which your body is resetting, balancing and restoring.

I often tell patients who are experiencing significant stress reactions that the first thing to do is work on achieving a high quality sleep pattern. That includes sleeping between seven and eight hours per night and having longer periods of deep sleep cycles.

I was treating a middle-aged female patient who suffered from ongoing fatigue. We had ruled out major medical causes and I was pretty certain that her fatigue was caused by poor sleep quality. We did a comprehensive assessment of her sleeping environment and her sleep hygiene; I discovered that there was a street light right outside her bedroom window which let the light stream directly into my patient's bedroom. Obviously, she couldn't do much about the street light but she could address the window treatments. She purchased inexpensive blackout draperies for her window and this one change significantly improved the quality of her sleep. She started waking up feeling more refreshed and within a relatively short period of time, her daily energy level improved as well. There is no substitute for quality sleep.

Here is a list of what we refer to in our practice as sleep hygiene techniques designed to help you achieve optimal sleep patterns.

1. Eat a healthy diet consistently throughout the day to maintain balanced blood sugar levels.

2. Sleep in complete darkness. There should be no ambient light from a night-light, bathroom and hallway or outside.

3. Set a comfortable temperature in your sleeping area—a room that is too cold or too hot can affect sleep quality.

4. Turn off electronics including smartphones, TVs, iPads, etc., at least one hour before bed.

5. Stop working at least one hour before bed.

6. When you get into bed and are ready to sleep, in your mind create a list of the moments during the day when you felt best. I call this a subconscious greeting for sleep. This makes you run through your day looking for all the good instead of focusing on what you did wrong, what you shouldn't have said, etc.

POSITIVE MENTAL ATTITUDE

1. **Attitude of gratitude.** When you live every day in a state of appreciation for all that you have, instead of focusing on the absence of what you would like to have, your mind will undoubtedly be in another emotional state. Everyone goes through tough times, but constantly thinking about what makes you unhappy or what you don't yet have continues to activate a state of negative emotion. Each morning gives you a new chance to start fresh. In your mind run through a list of what is going well for you as you're getting ready for the day. It may be as simple as being thankful for your running water, your hot shower, your loving pet, your cozy slippers or your morning cup of tea or coffee. The

list can be endless, but it requires being cognizant. Doing so will let you stay in a more peaceful
emotional state.

2. **Gratitude journal.** Although you can go over
   in your mind all the things in your life you are
   grateful for, actually writing down these items tends
   to amplify the positive emotions. Your gratitude
   journal can be as simple as a lined school note-
   book, or you can purchase an attractive book
   filled with beautiful illustrations. (These types of
   notebooks make a wonderful gift.)And of course,
   you must write in it on a regular basis; every day
   is ideal. It's important to think about why you are
   grateful for each item as your write in your journal.
   To avoid repetition, I like to suggest writing three
   new things each time, along with seven others that
   could be repeats (good coffee is a frequent repeat
   for me). You'll notice how different you feel after
   just two weeks of daily gratitude journal entries.

3. **Time outside in nature.** Going outside and
   walking around, without any specific purpose other
   than just being outside, is what I like to label as
   green light therapy. Noticing the pure beauty in
   nature, which plants are blooming, which species
   of birds are around, how the clouds appear, the
   sounds of little critters and how the light filters
   through the tree limbs is the essence of enjoying
   the natural abundance that surrounds you.

4. **Treat yourself well.** I see so many people in my
   life and clinical practice who treat themselves in
   ways they would never treat any of their loved
   ones. Unfortunately, many of these people are
   mothers who are putting themselves last on the
   list. Of course, the selflessness of caring for all the
   needs of a family is a beautiful thing, but it often
   leaves mothers running on empty. By making the
   choice to be kind to yourself, you can take away
   much of the perceived stress that is self-created.
   I often tell my patients and even some of my
   friends, "Don't be so hard on yourself. Give that
   girl a break!"

Managing the Stress Response

Rarely do we consciously choose the stressors that are in our
lives, but we can choose how we respond to those stressors.
And if we meditate on a regular basis, we know our reactions
that link up fear and anxiety to this response are significantly
altered. "Stress is being here and wanting to be over there." I
love this quote from Eckhart Tolle, one of today's great spiritual teachers. According to Tolle, if we could all live from
a place of peace in the present moment, no matter what is
happening, then we wouldn't experience stress. (And you
wouldn't be reading this section!) While this view sounds
ideal, the reality is that it's not easy to avoid stress. There is
already a library of literature on stress, its impact on your
life, techniques to avoid stress and more, so I will only say

that I firmly believe that in order to get to a place where you no longer react to all the stressors in your life, you should start a regular practice of meditation. (Okay, now just skip to the section on meditation towards the end of the book!)

"When you change the way you look at things, the things you look at change." Dr. Wayne Dyer, one of my greatest teachers who had a profound impact on my life said this. It is so simple but very effective advice whenever you're feeling overwhelmed. Recognize that it is just that, a feeling and if you choose to think differently, your feelings will also change. I know this seems challenging when you are in the middle of a vicious cycle of negativity but it truly does work. I have used these techniques with my patients and on myself and have seen tremendous results. When you consider what is bothering you the most today, and fast forward one year, you may realize those matters are insignificant. It's a good tool that helps you step back from a situation and get a big picture perspective. Trust me, you will look back and wish you focused more on the people and things that bring you joy instead of what makes you feel ugh (for lack of a better term), and we all know how that feels.

Five-Minute Super Charge Techniques

Since you may have already exhausted the tools in your toolbox to calm your overreacting stress response system, I am going to share with you some of the easiest and most effective techniques to lessen the stress in your life. And you will see results quickly! I have used these in clinical practice with great success. Personally, I like creative visualization the most, but everyone has his or her own preferences so try all of them and decide what works best for you. Any of these strategies can help you turn down that volume stress button in only five minutes.

RELAXATION BREATHING

"Just stop and take a few deep breaths." This phrase may have been directed at you at some point in your life, or perhaps you were the one saying these words to a loved one. And for good reason, they work. Slowing down and focusing on your breathing is one of the easiest ways to relax and feel centered. As soon as you feel like those stress hormones are pumping, take action with deep breathing. You can do this anywhere—at your desk, a chair in the kitchen or before you go to bed—to get into a relaxed state. Here are the five simple steps:

1. Close your eyes.

2. Start to focus on your pattern of breathing as air flows in and flows out.

3. Now count to three as you breathe in, breathe out, counting to five.

4. Focus only on your breathing, and as thoughts come into your mind, release them and continue to count your breaths.

5. After five minutes, stretch and you will feel much calmer than when you started.

CREATIVE VISUALIZATION

As I mentioned, this technique is my favorite and is great for anyone who enjoys the creative process, and is particularly easy if you are a bit of a daydreamer. The key here is visualizing something that makes you feel good so you don't hold any negative emotion. For example, if you love the beach and you visualize everything about being there but you feel negative emotion about it because you can't afford a beach vacation, then you are defeating the purpose of this exercise. Find a setting or situation to visualize that only brings you joy or serenity.

1. Decide what you are going to think about, i.e., a social gathering, swimming in the ocean, walking in a field of wildflowers.

2. Set a timer for five minutes. (You can set it for a little longer, but not too long because your mind will tend to wander. The goal here is to stay focused on this positive feeling.) If you can, choose a ringtone that is pleasant, not startling.

3. Start your session, thinking about your happy place. It doesn't need to be the same thing every time, but going back to something familiar can be useful and induce those positive emotions you're seeking.

4. Visualize everything from the temperature, who is there with you, the lighting and the sounds you hear to what you are wearing. Use any and all details that enhance the "reality" of the setting.

5. When your timer rings, take a couple of deep breaths and open your eyes.

FULL BODY RELAXATION

If you feel like your shoulders are riding up close to your ears, your body could benefit from some directed muscle relaxation. The stress response can leave you feeling tense all over and one way to override this is to get a relaxing massage; but when you don't have the time (or money) to spend on regular massages, the next best thing is a segmental guided full body relaxation. This exercise can be done seated or laying down.

1. Get comfortable and close your eyes.

2. Starting at your toes and feet, tense your muscles there and hold for five to 10 seconds, then relax.

3. Move up to your lower legs, then your knees and upper legs. Tense for five to 10 seconds and release.

4. Tense the muscles in your pelvis and buttocks, hold five to 10 seconds and release.

5. Next, move to your hands, make a fist and hold five to 10 seconds, then relax.

6. Now move to your lower arms, then upper arms, tense five to 10 seconds and release.

7. Move to your abdomen, hold tension there for five to 10 seconds and release.

8. Up to your chest, tense your muscles for five to 10 seconds, relax and then take a deep breath in and out.

9. Go to your lower back, then upper back and repeat, again followed by a deep breath in and out.

10. Now move to your shoulders and neck, hold tension for five to10 seconds and release.

11. The last part of your body is your head and face. Tense up, followed by release.

12. Finish by tensing your entire body for 10 seconds and relax with a couple of deep breaths, feeling the deep level of relaxation all over. Big stretch and you are done!

EMOTIONAL FREEDOM TECHNIQUE (EFT)

My first exposure to EFT was quite some time ago when a friend of mine used this technique to get rid of his seasonal allergy symptoms. I was skeptical, but intrigued. I learned that EFT is essentially a series of acupuncture points that you tap lightly on repeatedly to help energy flow through meridians.It is thought to be more useful for emotional symptoms. In recent years, EFT has gained a strong following and is now commonly referred to as "tapping." Many people claim to have reduced stress and fewer negative emotions from tapping. If you are a more tactile person, this technique is worth a try. There are many free resources available online (watching a video is ideal) showing how to tap away your worries.

CHAPTER 16

BODY

Knowing some of the unique genomic factors that are part of your book of life allows you to personalize your program in the Body portion of the Precision Health Program and thereby reap the benefits of what was discussed in part 1. I explain to my patients, "This is where we can test, not guess." If you have already had a genomic wellness profile completed, then use the check boxes to customize your Body program. Even if you don't know your genomic profile, you probably have observed how your body reacts to certain combinations of foods and types of exercise so you may be able to answer these questions. However—and this is another reason you should have a genomic profile— your reaction to some items such as vitamins is impossible to discern without testing.

Regardless of whether you have had the test or not, this section is full of useful ways to optimize your diet to include foods that interact with your DNA. Science suggests that your epigenome is at least equally important as your genome. For best results, and to personalize your healthy lifestyle even further, consider having a genomic wellness assessment combined with genetic counseling from a healthcare provider who specializes in the field of precision medicine.

PHP: Body Checklist

- ▶ I have plenty of energy
- ▶ I maintain a healthy body composition
- ▶ I don't rely on food and alcohol to make me feel better
- ▶ I eat a whole foods-based diet, containing a wide variety of health-promoting foods
- ▶ I avoid genetically modified foods
- ▶ I avoid foods that my body reacts to
- ▶ I feel well after I eat, and between meals
- ▶ My digestive system is in good condition
- ▶ I understand the difference between good and bad carbs
- ▶ I tend to eat only lower glycemic-load foods
- ▶ I eat superfoods everyday
- ▶ I am free of pain
- ▶ I maintain a schedule of regular physical activity
- ▶ I have great posture
- ▶ My body is flexible

Nutrition

1. MATCHING DIET

Based on genetic testing, a dietary profile that matches your genetics can be determined. This dietary plan is based on the ratio of carbohydrates, fats and proteins. My husband and I were very fortunate to match up to the same Mediterranean style diet which is 20% protein, 35% fat and 45% carbohydrates.

- ▶ Mediterranean
- ▶ Low carb
- ▶ Low fat
- ▶ Balanced

Knowing which diet is your match determines the ratio of macronutrients. Once you know your desirable ratio, you need to consider macronutrient sources and food combinations. Your genomic profile will help you decide which super fats are right for you so you can check that off next.

2. FATS

My Super Fat(s):

- ▶ **Monounsaturated fats** Sources: olives, olive oil, avocado, certain nuts and nut butters (almonds, pistachios, peanuts), seeds (sesame, pumpkin, sunflower)
- ▶ **Polyunsaturated fats** Sources: fish, omega-3 eggs, evening primrose oil (an omega-6), flax seed (omega-3), chia seed (omega-3), walnuts

3. CHOOSING THE BEST CARBS

Most of you are probably familiar with the distinction between good carbs and bad carbs, which is determined by a measurement called glycemic load (GL). When you eat food containing carbs, such as pasta, how much your blood sugar rises in response to a specific amount consumed results in a GL value. You want to have a lower glycemic load. Starches like instant rice, processed cereals and flour products have the highest glycemic load, while starchy vegetables and some fruits (like bananas) have a moderate glycemic load, and non-starchy vegetables, beans and most fruits have the lowest glycemic load. No matter which matching diet you follow, the bulk of your meals will contain plant foods, when you take your oils and carbs into account.

It's also important to understand that a number of factors, including water content, preparation and shape, can influence the glycemic effect of foods in your diet. For example, pasta shapes with more surface area like spaghetti have a higher glycemic value than shapes like fettuccine. Also, if the pasta is cooked al dente it has a lower GL than overcooked pasta. The concept of food-combining alters the glycemic load; review following chart.

Food	Glycemic Load
Peanuts, 50g	0
Hummus, 50g	1
Soy beans, canned and drained 150g	1
Strawberries, fresh 120g	1

Food	Glycemic Load
Carrots, 80g	2
Cashews, 50g	3
Mixed Nuts, 50g	4
Milk, 250mL	4
Watermelon, 120g	4
Pear, 120g	4
Green peas, frozen, boiled 80g	4
Blueberries, fresh, wild, 100g	5
Peach, 120g	5
Apple, Golden Delicious 120g	6
Dark chocolate, 50g	6
Black beans, 150g	7
Wheat tortilla, 50g	8
Lentils, canned and drained 150g	9
Chickpeas, canned and drained 150g	9
Cherries, fresh, dark, 120g	9
Grapes, black, 120g	11
Corn tortilla, 50g	12
Oatmeal, 250g	13
Quinoa, 150g	13
Milk chocolate, 50g	13
Fettuccine, 180g	15

Food	Glycemic Load
Brown rice, 150g	16
Banana, 120g	16
Pretzels, baked, 30g	16
Corn on the Cob, 150g	20
Spaghetti, white, boiled 10 min, 180g	24
Bagel, white, 70g	24
Spaghetti, white, boiled 20 min, 180g	26
Raisins, 60g	28
Instant oatmeal, 250g	30
Macaroni and Cheese (Kraft), 180g	32
Corn pasta, 180g	32
Baked Russet potato, 150g	33
Rice pasta, 180g	35
Sweet potato, baked, 150g	42
White rice, 150g	43

4. FOOD COMBINATIONS

Usually, our meals and snacks contain a combination of carbs, fat and protein. The amounts of each can vary widely which allows you to use the concept of food combinations to your advantage. Consuming fats and proteins slows down the digestive process and reduces the glycemic load of the carbohydrates. Furthermore, fiber and water content also improves the glycemic load of the meal. Although an apple has a relatively low glycemic load, eating it with several tablespoons of almond butter (which contains healthy fats, some protein and fiber) will further lower the effect on your blood sugar. When you're preparing meals and snacks, always think about food combinations, which are a very simple way of improving your health. If you are going out with friends for dinner, which can often start with drinks, try to eat a small snack containing fat and protein before heading out to buffer the effect of the carbs. Alternatively, waiting until your meal comes and enjoying a glass of wine with your food is another good option.

5. PROTEIN

Sources of dietary protein widely vary. I am concerned with both the quality of the protein and that the quantity is a match for your genetic and metabolic needs. In general, most of the calories in your diet will come from plant sources. Protein can also be sources from plants, like organic soy and other legumes, but it is often difficult to achieve the levels of protein required for metabolism without adding a plant-based protein shake. The addition of branched chain amino acids (BCAA) to plant protein shakes like pea and rice is critical for achieving a high quality protein score. Sustainable sources of fish (not consumed daily because of the risk of mercury exposure), eggs and poultry are good options for protein, provided they are produced free of chemicals.

6. TIMING

I will say it again, what works well for one person might not work so well for another. Generally, eating within 30 minutes after getting up in the morning can be beneficial, but some people may do better compressing their calories into a smaller time frame during the day, a practice called "intermittent fasting."

Eating every two to three hours during the day, i.e., three meals and two to three snacks daily, may keep blood sugar more stable than eating one or two very large meals. The lowest glycemic foods should be consumed early in the day to take advantage of a phenomenon called "the second meal effect." The meal you have for breakfast will affect how you respond to lunch. For example, if you eat a bagel for breakfast and a green salad with grilled chicken for lunch, you will have a higher glucose level following that same lunch than if you had a plant-based protein shake for breakfast.

7. RESPONSE TO CERTAIN FOODS

▶ **Caffeine Sensitive, Slow Metabolizer** Limit consumption of coffee as much as possible. Increased coffee consumption increases your risk of having a heart attack.

▶ **Lactose Intolerant** (possibly)

1. If you carry a profile which indicates you might be lactose intolerant and you are experiencing gas, bloating, cramps, nausea, diarrhea anywhere between 30 minutes to two hours after eating, you can:

2. Eliminate dairy completely. If your tummy troubles go away, then it's likely you are lactose intolerant (although another possibility is that you have a dairy allergy).

3. Eliminate only certain dairy foods from your diet such as milk, ice cream and all cheeses except aged cheeses like cheddar, parmesan and Swiss. These contain only trace amounts of lactose. An easy way to check your favorite cheese for lactose is to look at the amount of sugar contained in the cheese, because lactose is a sugar. Cheeses with trace amounts of lactose will have zero sugar on the label. In general, aged cheeses are lower in lactose, while fresh, unripened cheeses such as mozzarella and cottage cheese are higher in lactose.

4. Take lactase pills every time you eat meals that contain dairy. Depending on the amount of lactose, the severity of your condition and even your age, you may benefit from using lactase pills if you are lactose intolerant.

▶ **Bitter Sensitive** Tips for including healthy bitter-tasting foods:

1. Prepare cruciferous vegetables like kale, cabbage, cauliflower and broccoli by roasting them in the oven with a healthy oil to improve the taste.

2. Add baby greens to a smoothie with berries to help mask the flavor.

3. It's okay to add a sprinkle of salt which will make these foods more tolerable as long as the overall diet is very low in processed foods, which is where most of the sodium intake in a typical diet is found, *and* the person is not sodium sensitive,

hypertensive or on a very low sodium diet prescribed by a physician for conditions such as chronic kidney disease or high blood pressure.

▶ **Sweet Taster** Tips for reducing intake of foods high in sugar:

Many people who are trying to lose weight often fail because they eat too much sugar and have trouble controlling their cravings. In order for your body to function, you need a constant supply of glucose in the bloodstream, so craving sweets makes sense. However, the seemingly constant sweet tooth and the general obesity epidemic aren't going away. You need to make healthier choices. Among the preferable options to satisfy your quest for sweet foods are:

1. Fresh fruits and berries: These are great snacks that have natural sugars plus fiber, which slow down the digestion of sugar. Also, these foods contain beneficial phytochemicals including powerful antioxidants that help combat environmental stress and fight the aging process.

2. Eat a small piece of intense dark chocolate instead of a whole milk chocolate candy bar. Choose chocolate that contains at least 70% cocoa solids, preferably organic. Eating a bite-size piece daily is actually good for your health. Cocoa contains potent antioxidants like those found in red wine. A piece of dark chocolate and a glass of red wine—what could be better?

3. Taking a high quality chromium supplement can help reduce your cravings for sugar and improve how your body handles glucose. Chromium can help reduce blood glucose, HbA1c levels and cholesterol.

Over time, sugar cravings decrease when you've been following a diet low in refined sugars and carbs. Usually after 30 to 60 days the taste for sweet foods changes and becomes less intense. In the meantime, follow the above steps and you should notice a difference.

▶ **Alcohol Flush** Sorry!! No solutions here. If you have this SNP, you really should avoid consuming alcohol, not just because of the flushing and uncomfortable feeling you get when you drink, but because of the more severe health consequences associated with this SNP when you continue to include alcohol as part of your lifestyle.

8. QUALITY OF FOOD SUPPLY

Unfortunately, in our food supply, all foods are no longer created equally. With the massive level of factory farming, including the genetic engineering of crops, and most recently, even salmon, the quality of our food has dramatically declined. However, on the positive side, there is an increasing trend by consumers demanding organically and sustainably produced foods, grown without harmful pesticides and non-GMO (genetically modified organisms) and prepared with more traditional methods. I always recommend buying organic because these foods have higher levels of those super phytochemicals that essentially "speak" to your DNA, thereby modifying the expression of your genes. It's also worthwhile buying locally grown produce whenever possible; these foods haven't been sitting on a truck or a store shelf, which can degrade the vitamin content. Look for a Community Supported Agriculture (CSA) in your area and try growing some of your own food.

9. SUPERFOODS

Over the years, this list has remained fairly static since these foods have been shown to promote health. Today, though, you can go a step beyond based on what the epigenome shows. Far ahead of this entire pack is one thing you can consume daily that exceeds the importance of every item on this list, and quantity matters. It is liquid, has zero calories... WATER. I recommend dividing your weight (in pounds) in half and drinking that many ounces, spread out over the course of a day. If you weigh 150 pounds, then drink 75 ounces of water daily.

- Chocolate
- Salmon, Cold Water Fish (Cod)
- Green tea
- Garlic
- Blueberries and other dark berries
- Turmeric
- Coffee
- Red Wine
- Broccoli
- Soy, non-GMO

10. NUTRITIONAL SUPPLEMENTS

Based on genomic testing, you may have increased need for specific nutrients that can be easily added in the form of daily supplements. I have taken dietary supplements for years, but it wasn't until I had genomic testing that I nailed down the foundational core of "extras" that I really needed to take on a daily basis. What a difference it makes! Ideally, you should work with a healthcare provider trained in functional medicine to decide which ones are right for you. The following is a list of the dietary supplements you should consider based on the nutrients that I covered in part 1. Wellness genomic testing related to vitamins continues to expand, but these are the nutrients for which there is the most evidence to date of their efficacy. I also take a complete multivitamin and mineral with a mixed phytochemical extract for foundational support and then add the targeted micronutrients for extra support.

- ▶ Vitamin A
- ▶ Folate, L-5-MTHF
- ▶ Vitamin C
- ▶ Vitamin B_6
- ▶ Vitamin D
- ▶ Vitamin B_{12}, Methylcobalamin
- ▶ Vitamin E
- ▶ Omega-3

11. DON'T FORGET TO FEED YOUR BACTERIA!

You remember from chapter 8 on microbiome that maintaining a gut full of healthy bacteria requires the right balance of macronutrients and consumption of certain foods that help the healthy bacteria thrive. Too much fat or protein is detrimental to gut health, as is a diet with highly processed carbohydrates and added sugars. The jury is still out on the exact ratio of fat to carb to protein that is ideal for the health of your gut bacteria. I advise to follow your whole foods-based, genetic matching diet, which will keep you in the zone of macronutrients that promote healthy gut bacterial balance. In addition, I also encourage eating the specific foods listed below.

When I order comprehensive GI testing on patients in my practice and find dysbiosis, or imbalance, in the gut flora, I often recommend a therapeutic course of care that involves a "microbiome dietary balancing plan." Sometimes, I start patients out on a liquid diet of specific gut-friendly fresh foods, giving the digestive system a rest while providing the best fuel sources for beneficial bacterial growth. Along with that I often advise strain-specific probiotics and prebiotic therapy. This combination is referred to as a *synbiotic*. Fiber is very important, somewhere in the neighborhood of 40-50 grams per day. You reach this level by eating a ton of non-starchy vegetables, especially the stalks, which are critical for good microbial growth. Make sure you are eating the whole vegetable and not just the asparagus tips or the broccoli florets. Foods to include regularly in your diet are:

Microbiota Friendly Foods

chocolate	garlic	leeks	chickpeas
asparagus	mushrooms	leafy greens	tomatoes
fermented foods: Kimchi, Miso	radishes	carrots	turmeric

Interestingly, eating fresh organic vegetables right out of the garden is also thought to provide some benefit. When I was a child, my brother and I had our own small vegetable garden where we grew carrots and green peas. Of course, we ate the peas directly off the plant, but I remember pulling up carrots and only giving them a quick rinse outside before devouring them. We also routinely ate wild strawberries, blueberries and blackberries (really any berry we knew was edible) right off the bushes and never got sick. Now, fast forward three decades later and I have a lovely garden right outside our front door filled with plants that my three kids will eat fresh out of the bed.

12. MAINTAINING A HEALTHY RELATIONSHIP WITH FOOD

In part 1 of this book on genomics, I reviewed six gene variants that influence your relationship with food. You may also remember that I said I have successfully used several strategies with my patients to improve their relationship with food. I have listed all of these techniques here, so if you have had your genomic profile, you can go through and check off which ones you can use. If you have not been tested, and you know you struggle with any of these issues, the same techniques can work for you, too.

▸ **Eating Disinhibition**

1. **Recognize your triggers** for emotional eating and when you see one coming have a substitute lined up. Find something other than food that elevates

your mood—music, exercise, dancing, talk to
that friend who always makes you laugh!

2. **Don't buy those foods that are your saboteurs**.

3. **Don't start eating the foods** that you know you
can't resist finishing. You really do need to practice avoidance.

4. **Measure or portion your foods**. You can't put a
large piece of chocolate cake on your plate and say
to yourself, "Oh, I'll just take a couple of bites."

5. **Tell your friends**, your partner, anyone who will
listen!! Get it out in the open and you will feel so
much relief. People can save you, especially in
social settings. Ask them to "keep me away from
that buffet table" or "don't offer me dessert."

▶ **Food Desire**

1. **Increase the pain/pleasure ratio**. This is really
the most precise technique to master and conquer
this eating behavior trait. You must make it much
more difficult to get your favorite foods. There
comes a point where the trouble expended to get
that food is just not worth it. You have to figure
out what is your threshold. You can get very
creative here. Make it a game. For example, you
might say you will only get ice cream if you have
to walk a half hour to and from the store where
you'll buy the ice cream.

2. **Eat enough protein and calories early in the
day**. Starting your day off with the right balance

of macronutrients, making sure you have at least 15 grams of protein at breakfast and consuming enough calories early in your day, is vital. Then, eat balanced meals frequently throughout the day to keep the hunger hormone called ghrelin in check. The way you eat during the day affects ghrelin levels. Ghrelin levels will rise if you don't eat enough calories during the day, which is why many people end up famished when they come home after work and then make poor choices for dinner and night snacks. Feeling hungry triggers the food desire behavior.

3. **Put the concept of nature vs. nurture to work for you**. Your genetics are not your destiny. This is a behavioral trait that is essentially based on making choices. We make choices all day long, every day. Make a choice to not go out of your way to get foods that don't nourish your body. I know this is one step that is easier said than done. It requires work, but the more you make these choices, the more you "practice," the better you get at it!

▶ **Sweet Tooth**

1. **Again, emphasize to family and friends** that this is how you are "wired." So don't hide how you feel. Own it, and say, "Yes, that's me. I have a sweet tooth" and tell everyone not to show their love for you with gifts of milk chocolate and jelly beans!

2. **Choose healthier sugars.** I have advised patients for many years even before I started testing them for these SNPs that when they feel like they want something sweet, try a small serving of frozen raspberries or a square or two of dark chocolate (>70% cacao).

3. **Use specific food-combining to your advantage.** The other way to handle your sweet tooth is from a physiologic standpoint by buffering the effect on your blood glucose/insulin secretion through a combination of carbohydrates with protein, fat and fiber. These food components lower the glycemic effect of a meal or snack, thereby diminishing a cascade that leads to fat deposition.

▸ **Satiety**

1. **Eat foods that are fiber rich.** This is the single most important thing you can do to deal with lack of satiety. I have often told my patients who complained of difficulty feeling full after an average-sized meal to drink an eight-ounce glass of water and eat an apple before dinner. The type of fiber found in an apple expands in your stomach with the water and provides a sensation of fullness, thereby reducing the amount of food you're likely to consume during the actual meal. You should also include fiber-rich foods with every meal.

2. **Eat at least every three hours.** Many of my patients find that by making sure they have a

steady stream of balanced meals throughout the day, they are less inclined to feel like they are running on empty all the time.

3. **Plan your personal menu for the day.** Knowing what you plan to have for your next meal can sometimes combat lack of satiety following a meal. I know many people who, when they are eating their meal, are thinking about what they are going to eat next. Writing out your meal plan can be a particularly effective strategy.

▶ **Snacking**

1. **Planning is key.** When you go grocery shopping, make sure you have included healthy snacks on your list. When you are driven to snack frequently and you want to maintain balance, you need to choose your foods wisely. This can actually work to your benefit. The more we learn about the interaction of the vital phytochemicals in foods, we understand there's a complex interaction, beyond simply fats, proteins, carbs, vitamins and minerals. In fact, these tiny plant chemicals speak to our cells and to our genes. These messages are what signal genes to turn on and off (which is the topic of part 2 of this book). When your snacks are loaded with these superfood chemicals, like EGCG in green tea or polyphenols in dark chocolate, your state of health will reflect that. So savor the moment and embrace those two squares of dark chocolate.

2. **Eat smaller more frequent meals to keep daily calories in check**. If you are a grazer, then you need to make sure the calories don't quickly add up. Of course, we know the picture is much more complex than just calories in versus calories out, but calories still do count.

3. **Be prepared and always carry healthy snacks**. When you tend to eat frequently, you need to prepare your snacks so you don't have to rely on unhealthy vending machine options. Some of my patients carry portable snacks like protein bars, nuts and apples. You should also make sure you are drinking water regularly.

Exercise

DESIGNING THE ULTIMATE EXERCISE PROGRAM FROM YOUR GENETIC PROFILE

For some, exercise is a part of their life because it makes them feel better and that is the ideal situation. However, for many others, exercise is a part of maintaining a healthy lifestyle or a method (some might say a chore) to attain specific benefits, like achieving and maintaining a healthy weight, building muscle strength and tone, enhancing sports performance/endurance, improving mood and sleep, stress reduction, boosting immune function and reducing the risk of heart disease, stroke, dementia and cancer. Whatever your reason, there is no doubt exercise is a foundational pillar for optimal health.

When I think about expression of our genes and the vast amounts of research on why exercise is so beneficial for human health, it all makes complete sense. Exercise acts as a key to unlock your genetic potential. It can turn on genes that promote health, and turn off genes that can cause dysfunction and disease. In designing exercise programs for my patients, I take into account their current health problems, their personal goals and, of course, their genomic test results. For instance, some people may have more difficulty with aerobic capacity, based on their profile, while others may have the SNP associated with increased fat deposition in muscle when they overdo strength training. I also look out for those with the Achilles tendinopathy SNP, especially among patients who do endurance training and run marathons.

**Before starting a new exercise program, it's wise to consult your physician first.

THE BASICS

Aerobic exercise and endurance training increase your heart rate and breathing rate. Monitoring your heart rate is key to assessing your heart rate *training zone*.

The basic equation to calculate your maximum heart rate is: 220 – your age = _____.

Then take that number and multiply by 50–70% to achieve a moderate heart rate training zone, or 70–85% for a vigorous training zone.

Examples of moderate activity (50–70%):
▶ Brisk walking
▶ General gardening
▶ Water aerobics
▶ Tennis (doubles)

Examples of vigorous activity (70–85%):
▸ Running, hiking
▸ Bicycling, spinning
▸ Intensive gardening
▸ Tennis (singles)

The general recommendation is a minimum of 150 minutes of *moderate* aerobic exercise per week; it is preferable to split it up, i.e., five 30-minute sessions are preferable to three 50-minute sessions. However, when my patients are beginning an exercise program, I suggest they fit in the exercise whenever they can so if they do three sessions instead of five, that's fine. Remember, you feel better when you make progress. As you move forward in your program and continue to train your cardiovascular system, then your level of endurance improves, so you could achieve similar health benefits from only 75 minutes of *vigorous* aerobic activity. To achieve more extensive results, you can work up to either 300 minutes of moderate activity or 150 minutes of vigorous aerobic activity. For people who have built up to a very high level of endurance and have been screened for preexisting health problems, they may continue to enhance their program with higher intensity interval training to achieve additional health benefits.

Strength training is the other form of exercise shown to provide additional advantages. As you may remember from part 1 of this book, there is a SNP that may result in added fat deposition. Therefore, depending on what goals a patient has for exercise combined with the medical reasons I am recommending exercise, the training programs can vary quite a bit. Again, the important message is that using genomics as part of the evaluation of current health issues like obesity or diabetes allows for personalization of an exercise program. What you're trying to achieve with your program may be completely different than what your partner's goals are for exercising. For some, the differences in the programs may be subtle, but for others the training schedule could be quite dissimilar.

Strength training can be mind-altering for many. Lifting weights can put some people in a different mental zone, much like an intense yoga or meditation session. The act of being present, focusing on the task at hand, turning off the outside world and feeling strong is somewhat like a power-trip, but in a positive way. The cascade of feel-good hormones and neurotransmitters released during a session can help explain the lift in mood that people often experience. And of course, there are the other physical health benefits including improved blood sugar control, better bone health and increased lean body mass, which equates to a higher metabolic rate.

Depending on your current level of exercise and health status, adding strength training to your program—like other lifestyle changes—should be individualized to your own needs and fitness goals. Benefits can typically be seen when adding weight training about three times per week, with the level of intensity determined by taking into account genomics, health goals and medical conditions. If you are just starting out, it's wise to initially use a personal trainer to customize a program for you. Alternatively, ask a friend or family member who is very knowledgeable about weight training to assist you.

Frame

CARING FOR YOUR FRAME

First let me state my bias: my husband is a chiropractor. Not only have I personally experienced the benefits of chiropractic care, physical medicine, acupuncture, massage and related modalities, I have also worked side by side with my husband in our clinical practice for more than 10 years. I have also referred countless patients to him who have had positive results from chiropractic care. I also attribute regular chiropractic care to getting me through three pregnancies without the level of body discomfort and pain that can often accompany a pregnancy.

If you have experienced back pain, hip pain, joint pain or muscle spasms and/or tightness, you know very well how this discomfort affects your mood, energy and productivity level. When your physical structure is misaligned, the signals from your nervous system, which impact every body system, can be interrupted and impair optimal function. Pain and discomfort are your body's thermostat or indicator telling you something is wrong. For some people, taking over-the-counter pain relievers has become a daily routine. However, these medications are not without risk and they only mask the problem without addressing the underlying cause. Approaching pain and joint dysfunction with a whole body approach to improve mobility of the joints, optimize muscle function and relieve inflammation is vital for your overall health. Incorporating stretching and flexibility exercises, as well as a regular exercise program, are all facets of caring for your frame. Eating a healthy diet and addressing stress in your life also are essential to maintaining a healthy musculo-skeletal system.

In my clinical practice, I often see patients who are suffering from cardio-metabolic diseases and other chronic conditions who are also in pain. It is the pain that can prevent them from starting an exercise program or even having the motivation to make changes in their diet, both of which are critical for reversing a disease state. For these people, I routinely prescribe chiropractic care while helping them slowly introduce lifestyle changes. It is very important to address chronic pain, which is essentially a chronic stressor. Stress can impact gene expression and emerging research is exploring the connection between chronic pain and epigenetics.

CHAPTER 17

SPIRIT

Have you ever wondered why in the phrase "mind, body and spirit" that spirit comes last? I sometimes think this expression is a metaphor for our society. In our fast-paced world, it's almost as if we have forgotten who we really are and our greater purpose in life. Taking the time and space to slow down and recognize the inherent value of feeling connected at a higher level and being an active participant in meaningful relationships with loved ones and your *self* is what I believe makes us feel complete. In fact, after many years of clinical practice, I decided it was important to offer a comfortable environment for people to meet as a group and explore ways to become more connected. Guess what we called it? Group Connect. We meet once a month in our warm and inviting reception area and discuss a range of topics including meditation, intention and self-love and even sound healing. In this section, I will discuss the importance of having a guiding life-view, the power of connection and service to others. In addition, I will explain various meditation methods that you should consider. As with the previous sections, you should first do a mini-self assessment by completing the checklist to determine areas in which you may need to focus.

PHG: Spirit Checklist

- ▶ I have a guiding purpose in my life
- ▶ I am an optimist
- ▶ I feel connected on a higher level
- ▶ I participate in meaningful relationships
- ▶ I feel there is more to me than just this body
- ▶ Enjoying everyday life is important to me
- ▶ I can spend time alone or with friends and still feel great
- ▶ I meditate every day for at least 15 minutes
- ▶ I routinely ask myself how I can be of service to others, without expecting anything in return

Life-View

There are actually quite a few research studies looking at the concept of life-view and the role it plays in your health. Life-view is defined as having broader lifelong goals that serve to direct and organize your day-to-day activities and what you value as having importance in your life. In simple terms, it's the reason why you get up every day and do what you do.

It may be that caring for your family is what guides you, or improving the lives of others in a more "professional" setting, or perhaps both. Whatever it is for you, it remains with you throughout the day. And it doesn't have to be complicated. You don't need to enroll in a 12-month program to find your "true" purpose in life. As I have said, it's as easy as simply doing what ultimately brings you joy. You will know you feel a sense of peace and flow when you are doing what you love.

Your attitude toward life, whether you generally think things tend to work out or the world is just out to get you, is also key to your well-being. In fact, a recent study showed that cynics end up having a threefold greater risk of developing Alzheimer's and dementia. Optimists, on the other hand, have been observed to have increased longevity. They tend to be more likely to exercise, experience lower levels of stress hormones and generally take better care of themselves. Characteristics of people 95 years and older include being optimistic, more outgoing and, in general, easygoing. They also say that laughter is important to them and they tend to have larger social networks (more on that next).

Service

Opportunities abound to help people in need. If you are already someone who routinely volunteers your time for a cause, you know how good it makes you feel. Random acts of kindness on a regular basis can have the same effect. Perhaps you offer to take another person's shopping cart back from the parking garage, or you drop a few extra items into the food donation bin at the grocery store. When you consistently ask yourself, "How can I be of service today?" without expecting anything in return, then you are in service for the greater good. I believe helping is a natural tendency and leads to a deeper sense of connection, not only within our communities but on a grander scale. While I think ultimately that's what matters most, there are also physical benefits from helping behaviors. A recent study of more than 4,000 people with heart disease demonstrated that those who spent time providing nonpaid assistance to family and friends outside of their households experienced fewer depressive symptoms compared to those who provided no assistance. Those in the "helping" group also decreased their chances of having another cardiac event or dying in the two years in which they were followed. This essentially proves that when you serve others you also serve yourself, and that is a good thing!

Connection

We are all connected at some level. Some people you feel more drawn to than others, obviously your partner, family members and your close friends. Have you ever met someone new and felt like you have known that person for years? Of course, it's easy to say that you both have similar views or traits. But sometimes that is not enough to explain it. I believe it is critical to your health and well-being to invest yourself in these meaningful relationships. Science shows us that maintaining positive relationships with a spouse, friends, family, colleagues, even neighbors is associated with better health and increased longevity. In fact, having social connections can increase your survival by 50%! Yes, this stuff has been studied extensively. We could technically classify few social connections as risky health behavior. In fact, research has shown it is:

- ▶ The equivalent of being an alcoholic or smoking 15 cigarettes/day!
- ▶ Worse than not exercising!
- ▶ Twice as harmful as obesity!

I am focused on connection to others, but feeling connected at a higher level is also a foundational piece of your overall state of well-being. Meditation is one of the most powerful tools that can help you get better connected and allow wellness to flow.

Meditation

Meditation is a practice that you can easily incorporate into your life. It should never feel like work and should leave you feeling better than when you started. As with any other healthy habit, once you start and find the approach that works for you, you will want to meditate on a regular basis. In recent years we have seen many CEOs of major companies admit they are aficionados of meditation. This is not surprising given the effects meditation has on the brain and behavior.

Meditation promotes a state of calmness and allows those who practice it to become more responsive in a productive way when reacting to a particular situation. Very basic anatomy of brain function explains this process: The neural connections among centers in the brain that control fear and anxiety and the ego are altered in a positive way in people who regularly mediate. In order to maintain these pathways, you must meditate on a regular basis.

On a more fundamental level, we also know that meditation is "getting into the gap," as described by Dr. Wayne Dyer. It is a way of deepening our connection with our true self and of connecting to the pure, positive energy of the Universe, or God, or whatever your belief system may be.

Although there are relatively few scientific studies about meditation compared to exercise, we know wonderful health benefits can be attained from meditating. When you're in a state of deep relaxation that can be attained by losing the focus on thought, and allowing thoughts to sort of "pass by," activity of the brain changes and you enter a theta wave state. The deepest brain state achieved by highly practiced meditators, like Tibetan monks, is the delta state, where they are still alert, but most people only experience delta state during one of our stages of sleep.

The actual practice of meditation can be carried out in a variety of ways; what matters is finding the method that you will do regularly and makes a positive impact on how you feel and perceive your life experience.

MINI-MEDITATION

This method of meditation is ideal for both beginners and experienced practitioners of Transcendental Meditation. I learned about this method from Eckhart Tolle, a teacher and influential spiritual leader. In his best-selling book, *The Power of Now*, Tolle explains that using this method means you are fully present in the moment. When you are going about your day there are many activities that are routine, such as brushing your teeth, putting away dishes or getting dressed. Instead of thinking about what you are going to have for lunch or that meeting with your son's teacher later that day, choose to be present.

Perhaps the best place to start is the process of washing your hands. If you have ever spent time with young children who are washing their hands, you will notice a perfect example of a mini-meditation. Two of my children are still quite young so I've been blessed with the opportunity to observe this in action many times. When I first learned about this method, I took my 6-year-old daughter to the movies and halfway through she needed to go to the restroom. While we stood at the sink, I watched her as she remarked, "Oh, look at the pretty pink soap." As she watched it slowly stream into her hand from the pump, she then proceeded to look at it on the palm of her hand and say, "Wow, it's so pearly." Then she started to rub it up over her fingers, one by one, and then did the same on the other hand. This was all before she even got to the water part. Next, she placed her hands under the water and started to rub them around while the soap took on a different form and got "all bubbly" as the slow-flowing water "tickled" her between her fingers. She squeezed her hands together under the water as it splashed around in the sink and she giggled at the scene she could create with just her hands and some water. Without rushing her, I allowed her to finish on her own time and get her paper towel and dry the water off her hands, crumpling up the towel, which then became a "basketball" for the trash can: "Score!" This experience has forever changed the way I look at washing my hands and I have passed this simple, yet perfect story along to many of my friends.

CLASSIC MEDITATION

Most people who practice meditation on a regular basis use this method. You may also hear meditation referred to as a practice and people will say "during my practice..." If you have never tried meditation, I suggest you start with a guided meditation, which is an audio track on which someone speaks first, typically very softly, and may suggest a mantra or intention you create for the session based on whatever you may be working on. This helps prime you to relax and start slowing down your constant flow of thoughts. One of the most common complaints I hear is that people have too much difficulty shutting down their thoughts, which is even more of a reason why they should be meditating!

I recommend you try meditating after you have been outside taking a walk or just enjoying the beauty of nature. I often refer to this as "green light therapy," which is a very effective way to reduce the stress response. Getting outside on a consistent basis, observing the splendor of your surroundings and feeling appreciative of everything around you is a powerful way to stay in balance and enjoy the present moment. When you're outside, your brain activity can be induced into a more relaxed state that is one step from the state achieved in meditation. You have probably heard someone say he is going to go for a walk outside to "clear his head." This is the reason that it actually works so well!

The beginner's guide includes these steps:

Step 1. Go outside, walk around and enjoy your natural environment for 10–15 minutes.

Step 2. Find a peaceful place to sit (this can be inside) preferably while you're in comfortable clothing (though this isn't a requirement). You can meditate at work on your break if that is the best time for you.

Step 3. Set a timer on your phone with a low-volume, calming ringtone. Ten minutes is a great start.

Step 4. Close your eyes and focus on your breathing.

Step 5. As thoughts come (and they will when you first start), don't focus on them and let them sort of pass by. You have not failed if you are still having thoughts. The key is to not dwell on them. When you think about needing to pick up your child later at school, don't continue on mulling about when you will then drop off the books at the library, swing by the ATM and what you are going to cook for dinner. Continue focused on breathing in a relaxed way.

Step 6. Before you know it, your "friendly" phone alarm will sound and you will have completed your first meditation!

21st CENTURY MEDITATION

In this digital age, meditation has also evolved. With the advent of smartphones, there's also an app for that! There are countless apps and programs for guided meditations where you can adjust the type of music, nature sounds, voice tone, timing, etc. There's even an app for meditation "skeptics." Because of the profusion of apps, I no longer recommend specific meditation apps to patients because personal preference should dictate which you use. If you are new to meditation, know that there are many apps to choose from and you may have to try several before finding one that is a good match. For some people, attending a group meditation class or training session can be of benefit. It really depends on your personality and learning style. Let me emphasize again that meditation is a practice, much like exercise. You can't simply do it once and expect the effects to continue long after you have finished. It typically takes three to four weeks for you to start noticing a difference in how you feel.

Acknowledgments

First and foremost, I would like to thank my husband, Mark, for his enthusiastic support throughout the course of writing this book. When you announce that you want to write a book and you have a clinical practice and a family, it takes a very special spouse to provide you with the time and support needed to undertake a project of this magnitude. From taking all three of our kids to multiple weekend hockey tournaments while I stayed at home to write to late nights staying up with me while I was working to the unwavering emotional support he provided. I love you so much and am forever grateful for our partnership, in our marriage, as parents and as business partners.

To my children, Ethan, Lilah, and Kate, thank you so much for understanding the importance of Mommy's work while I missed some of our family time together. Each one of you amazes me every day with your inner beauty. You inspire me to give more to our world and always do the best I can to help others feel better.

Professionally, I would like to first thank Dr. John Troup, Chief Science Officer of Metagenics, for supporting this book and our shared vision to expand the clinical practice of Precision Medicine. Thank you also to our collaborators on epigenetics at ErasmusAGE Rotterdam, the Netherlands, I would also like to thank Dr. Gabriella van Djik, and Dr. Taulant Muka for their contributions, as well as Jenna Troup for coordinating this effort. Next, I would like to thank my team of doctors and support staff at my practice. Thank you to Drs. Aylah Clark, Elizabeth Jensen, Sally Machin and Angela Varvaras, and Abby Owens, Joyce Sayre, Lorraine Raymond and Philia Christensen for holding down the fort while I spent time writing and traveling and for your personal support. A very special thank you to Dr. Aylah Clark for helping come up with the title of this book!

Of course, this book would not be a reality without the committed support of my literary agent, Nena Madonia Oshman, who believed in me and the concept of this book. I am so grateful to you for "taking a chance" on a first-time writer who didn't start out having a large platform. Thank you to my editor, Debra Englander; you were an absolute pleasure to work with.

The ultimate inspiration for this book comes from my patients. I thank each and every one of you. I have learned so much. You have taught me the true power of lifestyle medicine put into personal practice. So many of you are like close friends and family to me and I treasure the relationships we have maintained through the years. The countless times a patient has asked me, "why don't more people know about this?" was the impetus for writing this book so that I could reach many more people and let them know how effective lifestyle medicine can be in improving their daily lives.

About the Author

Dr. Stagg has always been passionate about science and health. She started out as a doctoral student in biochemistry at the University of Iowa. She left the field of research to pursue a career as a physician so that she could work directly with patients to help them improve their health. Dr. Stagg graduated with a doctorate in Naturopathic Medicine from Bastyr University in Seattle, Washington. She is a naturopathic physician and owner of the integrative specialty practice, Whole Health Associates, LLC. Dr. Stagg is a well known expert in precision medicine and regularly appears in the media. She has been featured on NBC, ABC and CBS. She is also an educator for other physicians and a public speaker on optimal health and wellness. She continues to see patients, conducts clinical research studies and trains physicians in personalized lifestyle medicine at her private practice in Connecticut. Dr. Stagg is happily married with three wonderful children, and has a deep understanding of the demands of today's family lifestyle.

Health Status Questionnaire (HSQ)

Date _____

Rate each of the following symptoms based upon your typical health profile for the past 14 days.

Point Scale

0   *Never or almost never* have the symptom
1   *Occasionally* have it, effect is *not severe*
2   *Occasionally* have it, effect is *severe*
3   *Frequently* have it, effect is not severe
4   *Frequently* have it, effect is severe

HEAD

_____ Headaches
_____ Faintness
_____ Dizziness
_____ Insomnia
_____ TOTAL

EYES

_____ Watery or itchy eyes

_____ Swollen, reddened or sticky eyelids

_____ Bags or dark circles under eyes

_____ Blurred or tunnel vision

_____ TOTAL

(Does not include near or far-sighted)

EARS

_____ Itchy ears

_____ Earaches, ear infections

_____ Drainage from ear

_____ Ringing in ears, hearing loss

_____ TOTAL

NOSE

_____ Stuffy nose

_____ Sinus problems

_____ Hay fever

_____ Sneezing Attacks

_____ Excessive mucus formation

_____ TOTAL

MOUTH & THROAT

_____ Chronic coughing

_____ Gagging, frequent need to clear throat

_____ Sore throat, hoarseness, loss of voice

_____ Swollen or discolored tongue, gums, lips

_____ Canker Sores

_____ TOTAL

SKIN

_____ Acne

_____ Hives, rashes, dry skin

_____ Hair loss

_____ Flushing, hot flashes

_____ Excessive sweating

_____ TOTAL

HEART

_____ Irregular or skipped heartbeat

_____ Rapid or pounding heartbeat

_____ Chest pain

_____ TOTAL

LUNGS

_____ Chest congestion

_____ Asthma, bronchitis

_____ Shortness of breath

_____ Difficulty breathing

_____ TOTAL

DIGESTIVE

_____ Nausea, vomiting

_____ Diarrhea

_____ Constipation

_____ Bloated feeling

_____ Belching, passing gas

_____ Heartburn

_____ Intestinal/stomach pain

_____ TOTAL

JOINTS & MUSCLES

_____ Pain or aches in joints

_____ Arthritis

_____ Stiffness or limitation of movement

_____ Feeling of weakness or tiredness

_____ TOTAL

WEIGHT

_____ Binge eating/drinking
_____ Craving certain foods
_____ Excessive weight
_____ Compulsive eating
_____ Water retention
_____ Underweight
_____ TOTAL

ENERGY & ACTIVITY

_____ Fatigue, sluggishness
_____ Apathy, lethargy
_____ Hyperactivity
_____ Restlessness
_____ TOTAL

EMOTIONS

_____ Mood swings
_____ Anxiety, fear, nervousness
_____ Anger, irritability, aggressiveness
_____ Depression
_____ TOTAL

MIND

_____ Poor memory

_____ Confusion, poor comprehension

_____ Poor concentration

_____ Poor physical coordination

_____ Difficulty in making decisions

_____ Stuttering or stammering

_____ Slurred speech

_____ Learning disabilities

_____ TOTAL

OTHER

_____ Frequent illness

_____ Frequent or urgent urination

_____ Genital itch

_____ TOTAL

_____ **GRAND TOTAL**

Sources

Chapter 1

Green Tea

Henning, S.M., P. Wang, C.L. Carpenter, and D. Heber. "Epigenetic effects of green tea polyphenols in cancer." *Epigenomics* 5, no. 6 (2013): 729–41. doi: 10.2217/epi.13.57.

Chapter 3

AJCN

Salbe, A.D., A. DelParigi, R.E. Pratley, A. Drewnowski, and P.A. Tataranni. "Taste preferences and body weight changes in an obesity-prone population." *Am J Clin Nutr* 79, no. 3 (2004): 372–8.

Study Related to rs1726866

Dotson, C.D., H.L. Shaw, B.D. Mitchell, S.D. Munger, and N.I. Steinle. "Variation in the gene TAS2R38 is associated with the eating behavior disinhibition in Old Order Amish women." *Appetite* 54, no. 1 (2010): 93–9. doi: 10.1016/j.appet.2009.09.011.

Study Related to ANKK1/DRD2 2007

Epstein, L.H., J.L. Temple, B.J. Neaderhiser, R.J. Salis, R.W. Erbe, and J.J. Leddy. "Food reinforcement, the dopamine D2 receptor genotype, and energy intake in obese and non-obese humans." *Behav Neurosci* 121, no. 5 (2007): 877–86. Erratum-ibid. 122, no. 1 (2008): 250.

2008 Canadian Study Related to Diabetes Young & Old Groups

Eny, K.M., T.M. Wolever, B. Fontaine-Bisson, and A. El-Sohemy. "Genetic variant in the glucose transporter type 2 is associated with higher intakes of sugars in two distinct populations." *Physiol Genomics* 13;33, no. 3 (2008): 355–60. doi: 10.1152/physiolgenomics.00148.2007.

2008 UK 3000 Kids Appetite Control Study

Wardle, J, S. Carnell, C.M. Haworth, I.S. Farooqi, S. O'Rahilly, and R. Plomin. "Obesity associated genetic variation in FTO is associated with diminished satiety." *J Clin Endocrinol Metab* 93, no. 9 (2008): 3640–3. doi: 10.1210/jc.2008-0472.

2007 Diabetes Study

de Krom, M., Y.T. van der Schouw, J. Hendriks, R.A. Ophoff, C.H. van Gils, R.P. Stolk, D.E. Grobbee, and R. Adan. "Common genetic variations in CCK, leptin, and leptin receptor genes are associated with specific human eating patterns." *Diabetes* 56, no. 1 (2007): 276–80.

Chapter 4

Weightlifting Program for Men

Orkunoglu-Suer, F.E., H. Gordish-Dressman, P.M. Clarkson, P.D. Thompson, T.J. Angelopoulos, P.M. Gordon, N.M. Moyna, L.S. Pescatello, P.S. Visich, R.F. Zoeller, B. Harmon, R.L. Seip, E.P. Hoffman, and J.M. Devaney. "INSIG2 gene polymorphism is associated with increased subcutaneous fat in women and poor response to resistance training in men." *BMC Med Genet* 9 (2008): 117. doi: 10.1186/1471-2350-9-117.

Heritage Study Endurance Exercise

Garenc, C., L. Pérusse, J. Bergeron, J. Gagnon, Y.C. Chagnon, I.B. Borecki, A.S. Leon, J.S. Skinner, J.H. Wilmore, D.C. Rao, and C. Bouchard. "Evidence of LPL gene-exercise interaction for body fat and LPL activity: the HERITAGE Family Study." *J Appl Physiol* 91, no. 3 (2001): 1334–40.

2008 Russian Bodybuilding Study

Druzhevskaya, A.M., I.I. Ahmetov, I.V. Astratenkova, and V.A. Rogozkin. "Association of the ACTN3 R577X polymorphism with power athlete status in Russians." *Eur J Appl Physiol* 103, no. 6 (2008): 631–4. doi: 10.1007/s00421-008-0763-1.

chapter 5

2007 Science Study

Frayling, T.M., N.J Timpson, M.N. Weedon, E. Zeggini, R.M. Freathy, C.M. Lindgren, J.R. Perry, *et al.* "A common variant in the FTO gene is associated with body mass index and predisposes to childhood and adult obesity." *Science* 316, no. 5826 (2007): 889–94.

2012 Mediterranean Diet

Ortega-Azorín C., J.V. Sorlí, E.M. Asensio, O. Coltell, M.Á. Martínez-González, J. Salas-Salvadó, M.I. Covas, *et al.* "Associations of the FTO rs9939609 and the MC4R rs17782313 polymorphisms with type 2 diabetes are modulated by diet, being higher when adherence to the Mediterranean diet pattern is low." *Cardiovasc Diabetol* 11 (2012): 137. doi: 10.1186/1475-2840-11-137.

2013 Breast Cancer

da Cunha, P.A., L.K. de Carlos Back, A.F. Sereia, C. Kubelka, M.C. Ribeiro, B.L. Fernandes, I.R. de Souza. "Interaction between obesity-related genes, FTO and MC4R, associated to an increase of breast cancer risk." *Mol Biol Rep* 40, no. 12 (2013): 6657–64. doi: 10.1007/s11033-013-2780-3.

Quebec Family Study

Loos, R.J., T. Rankinen, Y. Chagnon, A. Tremblay, L. Pérusse, and C. Bouchard. "Polymorphisms in the leptin and leptin receptor genes in relation to resting metabolic rate and respiratory quotient in the Québec Family Study." *Int J Obes* 30, no. 1 (2006): 183–90.

Explore Study

Scherwitz, L., and D. Kesten. "Seven eating styles linked to over-eating, overweight, and obesity." *Explore* 1, no. 5 (2005): 342–59.

Chapter 6

Stanford

Nelson, Mindy Dopler, Prakash Prabhakar, Venkateswarlu Kondragunta, Kenneth S. Kornman, and Christopher Gardner. "Genetic Phenotypes Predict Weight Loss Success: The Right Diet Does Matter." Oral presentation at the American Heart Association's Joint Conference—50th Cardiovascular Disease Epidemiology and Prevention and Nutrition, Physical Activity and Metabolism—2010, San Francisco, California, March 2–5, 2010.

British Journal of Sports Medicine

Malhotra A., T. Noakes, and S. Phinney. "It is time to bust the myth of physical inactivity and obesity: you cannot outrun a bad diet." *Br J Sports Med* 49, no. 15 (2015): 967–8. doi: 10.1136/bjsports-2015-094911.

Warodomwichit D., J. Shen, D.K. Arnett, M.Y. Tsai, E.K. Kabagambe, J.M. Peacock, J.E. Hixson, *et al.* "The monoun-saturated fatty acid intake modulates the effect of ADIPOQ polymorphisms on obesity." *Obesity* 17, no. 3 (2009): 510–17. doi:10.1038/oby.2008.583.

Smith, C.E., K.L. Tucker, D.K. Arnett, S.E. Noel, D. Corella, I.B. Borecki, M.F. Feitosa, *et al.* "Apolipoprotein A2 polymorphism interacts with intakes of dairy foods to influence body weight in 2 U.S. populations." *J Nutr* 143, no. 12 (2013): 1865–71. doi: 10.3945/jn.113.179051.

Corella, D., G. Peloso, D.K. Arnett, S. Demissie, L.A. Cupples, K. Tucker, C.Q. Lai, *et al.* "APOA2, dietary fat, and body mass index: replication of a gene-diet interaction in 3 independent populations." *Arch Intern Med* 169, no. 20 (2009): 1897–906. doi: 10.1001/archinternmed.2009.343.

Crous-Bou, M., G. Rennert, R. Salazar, F. Rodriguez-Moranta, H.S. Rennert, F. Lejbkowicz, L. Kopelovich, *et al.* "Genetic polymorphisms in fatty acid metabolism genes and colorectal cancer." *Mutagenesis* 27, no. 2 (2012): 169–76. doi: 10.1093/mutage/ger066.

PPARG rs1801282

Hsiao, T.J., and E. Lin. "The Pro12Ala polymorphism in the peroxisome proliferator-activated receptor gamma (PPARG) gene in relation to obesity and metabolic phenotypes in a Taiwanese population." *Endocrine* 48, no. 3 (2015): 786-93. doi: 10.1007/s12020-014-0407-7.

Chapter 7

Finnish Study

Ferrucci, L., J.R. Perry, A. Matteini, M. Perola, T. Tanaka, K. Silander, N. Rice, *et al.* "Common variation in the beta-carotene 15,15'-monooxygenase 1 gene affects circulating levels of carotenoids: a genome-wide association study." *Am J Hum Genet* 84, no. 2 (2009): 123–33. doi: 10.1016/j.ajhg.2008.12.019.

2008 B6

Morris, M.S., M.F. Picciano, P.F. Jacques, and J. Selhub. "Plasma pyridoxal 5'-phosphate in the US population: the National Health and Nutrition Examination Survey, 2003–2004." *Am J Clin Nutr* 87, no. 5 (2008): 1446–54.

Chapter 8

Risperidone

Bahr, S.M., B.J. Weidemann, A.N. Castro, J.W. Walsh, O. deLeon, C.M. Burnett, N.A. Pearson, et al. "Risperidone-induced weight gain is mediated through shifts in the gut microbiome and suppression of energy expenditure." *EBioMedicine* 2, no. 11 (2015): 1725-34.

Bahr, S.M., B.C. Tyler, N. Wooldridge, B.D. Butcher, T.L. Burns, L.M. Teesch, C.L. Oltman, et al. "Use of the second-generation antipsychotic, risperidone, and secondary weight gain are associated with an altered gut microbiota in children." *Transl Psychiatry* 5 (2015): e652. doi: 10.1038/tp.2015.135.

2015 Study World Diabetes

Fallucca, F., L. Fontana, S. Fallucca, and M. Pianesi. "Gut microbiota and Ma-Pi 2 macrobiotic diet in the treatment of type 2 diabetes." *World J Diabetes* 6, no. 3 (2015): 403–11. doi: 10.4239 /wjd.v6.i3.403.

Chapter 9

Minnesota Family Study

Segal, N.L. *Born Together-Reared Apart: The Landmark Minnesota Twin Study*. Cambridge, Massachusetts: Harvard University Press, 2012.

ᴄʜapter 10

Dutch Famine Study

Roseboom, T., S. de Rooij, and R. Painter. "The Dutch famine and its long-term consequences for adult health." *Early Hum Dev* 82, no. 8 (2006): 485–91.

1997 Animal Study

Liu, D., J. Diorio, B. Tannenbaum, C. Caldji, D. Francis, A. Freedman, S. Sharma, *et al.* "Maternal care, hippocampal glucocorticoid receptors, and hypothalamic-pituitary-adrenal responses to stress." *Science* 277, no. 5332 (1997): 1659–62.

2004 Reduced Stress

Weaver, I.C., N. Cervoni, F.A. Champagne, A.C. D'Alessio, S. Sharma, J.R. Seckl, S. Dymov, *et al.* "Epigenetic programming by maternal behavior." *Nat Neurosci* 7, no. 8 (2004): 847–54.

ACE Study

Felitti, V.J., R.F. Anda, D. Nordenberg, D.F. Williamson, A.M. Spitz, V. Edwards, M.P. Koss, *et al.* "Relationship of childhood abuse and household dysfunction to many of the leading causes of death in adults. The Adverse Childhood Experiences (ACE) Study." *Am J Prev Med* 14, no. 4 (1998): 245–58.

McGill

Borghol, N., M. Suderman, W. McArdle, A. Racine, M. Hallett, M. Pembrey, C. Hertzman, *et al.* "Associations with early-life socio-economic position in adult DNA methylation." *Int J Epidemiol* 41, no. 1 (2012): 62–74. doi: 10.1093/ije/dyr147.

Naumova, O.Y., M. Lee, R. Koposov, M. Szyf, M. Dozier, and E.L. Grigorenko. "Differential patterns of whole-genome DNA methylation in institutionalized children and children raised by their biological parents." *Dev Psychopathol* 24, no. 1 (2012): 143–55. doi: 10.1017/S0954579411000605.

Chapter 11

Agouti Mice

Jirtle, R.L. "The Agouti mouse: a biosensor for environmental epigenomics studies investigating the developmental origins of health and disease." *Epigenomics* 6, no. 5 (2014): 447–50. doi: 10.2217/epi.14.58.

Swedish Study Bygren

Kaati, G., L.O. Bygren, M. Pembrey, and M. Sjostrom. "Transgenerational response to nutrition, early life circumstances and longevity." *Eur J Hum Genetics* 15, no. 7 (2007): 784–90.

Spain Chocolate

Crescenti, A., R. Solà, R.M. Valls, A. Caimari, J.M. Del Bas, A. Anguera, N. Anglés, *et al.* "Cocoa Consumption Alters the Global DNA Methylation of Peripheral Leukocytes in Humans with Cardiovascular Disease Risk Factors: A Randomized Controlled Trial." *PLOS One* 8, no. 6 (2013): e65744.

Chapter 12

Twin Studies

Fraga, M.F., E. Ballestar, M.F. Paz, S. Ropero, F. Setien, M.L. Ballestar, D. Heine-Suñer, *et al.* "Epigenetic differences arise during the lifetime of monozygotic twins." *PNAS* 102, no. 30 (2005): 10604–9.

Boston Omega 3

Ma, Y., C.E. Smith, C.Q. Lai, M.R. Irvin, L.D. Parnell, Y.C. Lee, L.D. Pham, *et al.* "The effects of omega-3 polyunsaturated fatty acids and genetic variants on methylation levels of the interleukin-6 gene promoter." *Mol Nutr Food Res* 60, no. 2 (2016): 410–9. doi: 10.1002/mnfr.201500436.

✑hapter 13

Nature

Sen, A., N. Heredia, M.C. Senut, S. Land, K. Hollocher, X. Lu, M.O. Dereski, *et al.* "Multigenerational epigenetic inheritance in humans: DNA methylation changes associated with maternal exposure to lead can be transmitted to the grandchildren." *Sci Rep* 5 (2015): 14466. doi: 10.1038/srep14466.

✑hapter 14

Geisel School Baby Formula

Madan, J.C., A.G. Hoen, S.N. Lundgren, S.F. Farzan, K.L. Cottingham, H.G. Morrison, M.L. Sogin, *et al.* "Association of Cesarean Delivery and Formula Supplementation With the Intestinal Microbiome of 6-Week-Old Infants." *JAMA Pediatr* 170, no. 3 (2016): 212–9. doi: 10.1001/jamapediatrics.2015.3732.

European Studies

Zaura, E., B.W. Brandt, M.J. Teixeira de Mattos, M.J. Buijs, M.P. Caspers, M.U. Rashid, A. Weintraub, *et al.* "Same Exposure but Two Radically Different Responses to Antibiotics: Resilience of the Salivary Microbiome versus Long-Term Microbial Shifts in Feces." *mBio* 6, no. 6 (2015): e01693-15. doi: 10.1128 /mBio.01693-15.

Examining 20 Different Diets

Haro, C., M. Montes-Borrego, O.A. Rangel-Zúñiga, J.F. Alcalá-Díaz, F. Gómez-Delgado, P. Pérez-Martínez, J. Delgado-Lista, *et al.* "Two Healthy Diets Modulate Gut Microbial Community Improving Insulin Sensitivity in a Human Obese Population." *J Clin Endocrinol Metab* 101, no. 1 (2016): 233–42. doi: 10.1210/ jc.2015-3351.

Cork Rugby Study

Clarke, S.F., E.F. Murphy, O. O'Sullivan, A.J. Lucey, M. Humphreys, A. Hogan, P. Hayes, *et al.* "Exercise and associated dietary extremes impact on gut microbial diversity." *Gut* 63, no. 12 (2014): 1913–20. doi: 10.1136/gutjnl-2013-306541.

Maternal Stress

Jašarevic, E., C.L. Howerton, C.D. Howard, and T.L. Bale. "Alterations in the Vaginal Microbiome by Maternal Stress Are Associated With Metabolic Reprogramming of the Offspring Gut and Brain." *Endocrinology* 156, no. 9 (2015): 3265–76. doi: 10.1210/en.2015-1177.

Chapter 16

Glycemic Load Table

Atkinson, F.S., K. Foster-Powell, and J.C. Brand-Miller. "International tables of glycemic index and glycemic load values." 2008 December. *Diabetes Care* 31, no. 12 (2008): 2281–83.

Chapter 17

Study of Cynics

Neuvonen, E., M. Rusanen, A. Solomon, T. Ngandu, T. Laatikainen, H. Soininen, M. Kivipelto, *et al.* "Late-life cynical distrust, risk of incident dementia, and mortality in a population-based cohort." *Neurology* 82, no. 24 (2014): 2205–12. doi: 10.1212/WNL.0000000000000528.

4,000 People with Heart Disease: Social Connections

Heisler, M., H. Choi, J.D. Piette, A. Rosland, K.M. Langa, and S. Brown. "Adults with cardiovascular disease who help others: a prospective study of health outcomes." *J Behav Med* 36, no. 2 (2013): 199–211. doi: 10.1007/s10865-012-9414-4.